MASS SPECTROMETRY OF PROTEIN INTERACTIONS

THE WILEY BICENTENNIAL–KNOWLEDGE FOR GENERATIONS

*E*ach generation has its unique needs and aspirations. When Charles Wiley first opened his small printing shop in lower Manhattan in 1807, it was a generation of boundless potential searching for an identity. And we were there, helping to define a new American literary tradition. Over half a century later, in the midst of the Second Industrial Revolution, it was a generation focused on building the future. Once again, we were there, supplying the critical scientific, technical, and engineering knowledge that helped frame the world. Throughout the 20th Century, and into the new millennium, nations began to reach out beyond their own borders and a new international community was born. Wiley was there, expanding its operations around the world to enable a global exchange of ideas, opinions, and know-how.

For 200 years, Wiley has been an integral part of each generation's journey, enabling the flow of information and understanding necessary to meet their needs and fulfill their aspirations. Today, bold new technologies are changing the way we live and learn. Wiley will be there, providing you the must-have knowledge you need to imagine new worlds, new possibilities, and new opportunities.

Generations come and go, but you can always count on Wiley to provide you the knowledge you need, when and where you need it!

WILLIAM J. PESCE
PRESIDENT AND CHIEF EXECUTIVE OFFICER

PETER BOOTH WILEY
CHAIRMAN OF THE BOARD

MASS SPECTROMETRY OF PROTEIN INTERACTIONS

Edited by

KEVIN M. DOWNARD
The University of Sydney
Sydney, Australia

WILEY-INTERSCIENCE

A JOHN WILEY & SONS, INC., PUBLICATION

Library of Congress Cataloging-in-Publication Data:

Mass spectrometry of protein interactions / [edited by] Kevin M. Downard.
    p. ; cm.
  Includes bibliographical references and index.
  ISBN 978-0-471-79373-1 (cloth)
  1. Protein-protein interactions.   2. Mass spectrometry.   I. Downard, K. (Kevin)
  [DNLM: 1. Proteins–analysis.   2. Mass Spectrometry–methods.   3. Proteins–
metabolism. QU 55 M414 2007]
  QP551.5.M37 2007
  572'.64–dc22   2006100244

10 9 8 7 6 5 4 3 2 1

CONTENTS

Preface   ix

Contributors   xi

**1 Direct Characterization of Protein Complexes by Electrospray
    Ionization Mass Spectrometry and Ion Mobility Analysis**   1
Joseph A. Loo and Catherine S. Kaddis

1.1 I ntroduction, 2
      1.1.1 Historical Perspective of ESI-MS for Measuring Protein
              Complexes, 3
      1.1.2 Types of Interactions that Are Probed by ESI-MS, 6
1.2 Critical Aspects of the Experimental Procedure, 8
      1.2.1 I nstrumental Parameters, 8
              1.2.1.1 Electrospray Ionization Source, 9
              1.2.1.2 Atmosphere/Vacuum Interface and Pressure, 9
              1.2.1.3 Mass Spectrometry Analyzers, 10
              1.2.1.4 Ion Mobility Analyzers, 12
      1.2.2 S ample Preparation, 1 5
1.3 Solution Phase Equilibria and Gas Phase Dissociation, 16
      1.3.1 Measuring Solution Dissociation Constants, 16
      1.3.2 Tandem Mass Spectrometry of Protein Complexes, 16
1.4 Conc lusions, 1 8
      Acknowledgments, 1 9
      References, 1 9

**2 Softly, Softly—Detection of Protein Complexes
by Matrix-Assisted Laser Desorption Ionization
Mass Spectrometry**   **25**

Kevin M. Downard

2.1 I ntroduction, 25
2.2 First Glimpses and the First-Shot Phenomenon, 28
2.3 Matrix and Solution Criteria to Preserve Protein Complexes, 30
2.4 Laser Fluence, Wavelength, and Ion Extraction, 32
2.5 Preservation of Protein Complexes on Conventional MALDI
     Targets, 3 5
2.6 A ffinity Targets and Surfaces Coupled to MALDI, 37
2.7 Conc lusions, 3 9
     References, 3 9

**3 Probing Protein Interactions Using Hydrogen–Deuterium
Exchange Mass Spectrometry**   **45**

David D. Weis, Suma Kaveti, Yan Wu, and John R. Engen

3.1 I ntroduction, 4 6
3.2 Hydrogen Exchange Background, 46
3.3 General HX-MS Method, 47
     3.3.1 Location Information Provided by HX-MS, 49
     3.3.2 Revealing Interactions by Comparison, 50
3.4 Interactions of Proteins, 50
3.5 Ex amples, 5 2
     3.5.1 Conformational Changes of Proteins During Binding, 52
     3.5.2 P rotein–Protein Interactions, 5 2
     3.5.3 P rotein–Peptide Interactions, 54
     3.5.4 Protein–Small Molecule Interactions, 55
3.6 Conc lusions, 5 7
     Acknowledgements, 5 7
     References, 5 7

4 Limited Proteolysis Mass Spectrometry of Protein Complexes   63

Maria Monti and Piero Pucci

4.1 I ntroduction, 6 3
4.2 Limited Proteolysis Analysis, 64
4.3 Ex perimental Design, 6 7
4.4 Probing Protein–Protein Interactions, 69
4.5 Probing Protein–Nucleic Acid Interactions, 72
4.6 Probing Protein–Ligand Interactions, 74

4.7 Probing Amyloid Fibril Core, 76
4.8 Conc lusions, 7 8
     References, 7 8

5 **Chemical Cross-Linking and Mass Spectrometry
   for Investigation of Protein–Protein Interactions   83**
   Andrea Sinz

   5.1 I ntroduction, 8 4
   5.2 C ross-Linking Strategies, 8 5
       5.2.1 B ottom–Up Approach, 8 5
       5.2.2 T op–Down Approach, 8 8
   5.3 Functional Groups of Cross-Linking Reagents:
        Reactivities, 8 9
       5.3.1 A mine-Reactive Cross-Linkers, 8 9
       5.3.2 S ulfhydryl-Reactive Cross-Linkers, 9 1
       5.3.3 P hotoreactive Cross-Linkers, 9 1
   5.4 C ross-Linker Design, 92
       5.4.1 H omobifunctional Cross-Linkers, 92
       5.4.2 Het erobifunctional Cross-Linkers, 93
       5.4.3 Z ero-Length Cross-Linkers, 9 3
       5.4.4 T rifunctional Cross-Linkers, 9 3
   5.5 Mass Spectrometric Analysis of Cross-Linked Products, 94
       5.5.1 Bottom–Up Analysis by MALDI-MS, 94
       5.5.2 Bottom–Up Analysis by ESI-MS (LC/MS), 94
       5.5.3 Bottom–Up and Top–Down Analysis
              by ESI-FTICR-MS, 95
   5.6 I dentification of Cross-Linked Products, 97
   5.7 Computer Software for Data Analysis, 99
   5.8 Conclusions and Perspectives, 99
       Abbreviations, 1 00
       Acknowledgments, 1 00
       References, 1 01

6 **Genesis and Application of Radical Probe Mass Spectrometry
   (RP-MS) to Study Protein Interactions   109**
   Simin D. Maleknia and Kevin M. Downard

   6.1 Genesis of Radical Probe Mass Spectrometry, 110
   6.2 The Reactive Residue Side Chains, 111
   6.3 Conditions Important to Radical Probe Mass Spectrometry
        Experiments, 1 15

6.4 Generation of Radicals on Millisecond Timescales, 117
6.5 Applications of RP-MS to Studies of Protein Interactions, 119
    6.5.1 I ntramolecular Interactions, 1 20
    6.5.2 Intermolecular Interactions: Protein–Peptide and
        Protein–Protein Complexes, 122
6.6 Onset of Oxidative Damage and Its Implications for Protein
    Interactions, 1 26
6.7 Application of Radical Oxidation to Study Protein
    Assemblies, 1 28
6.8 Modeling Protein Complexes with Data from
    RP-MS Experiments, 129
6.9 Conc lusions, 1 30
    References, 1 31

Index   **135**

PREFACE

THROUGH THE LOOKING GLASS —PROTEIN INTERACTIONS AS VIEWED BY MASS SPECTROMETRY

Mass spectrometry has come a long way from its role in the discovery of isotopes for many of the chemical elements. In just a few decades, difficulties with introducing large, highly polar molecules such as proteins into a mass spectrometer have been overcome and the mass spectrometer, in its many guises, stands as a central technology for the analysis and sequencing of proteins. Perhaps even more astounding, given its construct, is the increasing role that mass spectrometry now plays in the study of protein and other macromolecular interactions.

There are a large number of experimental approaches at hand with which to examine some facet of protein interactions. Although mass spectrometry is not yet routinely practiced by all researchers for this application, it nonetheless can provide a unique window into the nature and stability of these interactions. Developments on a number of fronts from the direct detection of protein complexes and assemblies, to the use of hydrogen isotopic exchange and other chemical labeling approaches with mass spectrometry, to the application of ion mobility mass spectrometry, and the preservation of protein complexes on activated surfaces, have all advanced the study of protein interactions by mass spectrometry. Importantly, the salient features of mass spectrometric analysis—namely, the ability to detect molecules at low sample levels, to do so in complex mixtures without their purification, and to perform the analysis rapidly—are all transposed to these studies.

The use of mass spectrometry to investigate protein interactions using any one individual approach or a combination of approaches is beginning to move from the domain of specialist research laboratories involved in their development to protein scientists and biologists in general. Over a decade on from the earliest observations, an appropriate juncture has been reached at which to review the progress made thus far and report on the latest discoveries and applications as well as new and ongoing challenges. At the time of preparation, there is no book available that covers these developments in a single authoritative volume. This book aims to bring together a series of chapters covering the many avenues with which to study protein interactions by mass spectrometry, each written by international authorities, and in some cases pioneers of the approaches.

In teaching students of the wonders and wherefores of mass spectrometry, I have likened the mass spectrometer to a well-trained dog. Largely obedient, quick to perform, precise in execution, the mass spectrometer eagerly, expeditiously, and expertly can analyze and sequence proteins. But as owners, or should I say custodians, we in the mass spectrometry research community would like our "dog" to jump a little higher, run a little faster, and not without a little satisfaction outperform other "animals" (read analytical technologies).

I am reminded of the words of Lewis Carroll from *Alice's Adventures in Wonderland.*

> *Will you walk a little faster? said a whiting to a snail,*
> *There's a porpoise close behind us and he's treading on my tail.*
> *See how eagerly the lobsters and the turtles all advance!*
> *They are waiting on the shingle—will you come and join the dance?*
> *Will you, won't you, will you, won't you, will you join the dance?*

The contents of this book allows one to peer through the looking glass to view the present state-of-play, presents the latest achievements and challenges, and leaves the reader to wonder about what might be possible in the years ahead. On behalf of the contributing authors, I invite you, the reader, to come and join the dance.

KEVIN M. DOWNARD

The University of Sydney
Sydney, Australia

CONTRIBUTORS

*Kevin Downard   School o f M olecular a nd M icrobial B iosciences, T he University of Sydney, Australia [email: kdownard@usyd.edu.au]

*John R. Engen   Department of Chemistry, University of New Mexico, Albuquerque, NM 87131-0001 [email: engen@unm.edu]

Suma Kaveti   Lerner Re search I nstitute, C leveland C linic F oundation, Cleveland, OH 44195

Catherine S. Kaddis   David Geffen School of Medicine, University of California, Los Angeles, CA 90095

*Joseph A. Loo   Molecular Biology Institute, University of California, Los Angeles, CA 90095 [email: JLoo@chem.ucla.edu]

*Simin D. Maleknia   School of Biological, Earth and Environmental Sciences, U niversity o f N ew S outh W ales, A ustralia [e mail: s .maleknia@ unsw.edu.au]

Maria Monti   Università di Napoli "Federico ll," Napoli, Italy

*Piero Pucci   Università d i N apoli " Federico l l," N apoli, I taly [e mail: pucci@unina.it]

*Andrea Sinz   Institute of Pharmacy, Martin Luther University, Halle-Wittenberg, Germany [email: andrea.sinz@pharmazie.uni-halle.de]

*Indicates corresponding author.

David D. Weis   Department of Chemistry, University of New Mexico, Albuquerque, NM 87131-0001

Yan Wu   Department of Chemistry, University of New Mexico, Albuquerque, NM 87131-0001

1

DIRECT CHARACTERIZATION OF PROTEIN COMPLEXES BY ELECTROSPRAY IONIZATION MASS SPECTROMETRY AND ION MOBILITY ANALYSIS

JOSEPH A. LOO

Departments of Chemistry and Biochemistry, and Biological Chemistry, David Geffen School of Medicine, University of California, Los Angeles, California 90095

CATHERINE S. KADDIS

Department of Chemistry and Biochemistry, David Geffen School of Medicine, University of California, Los Angeles, California 90095

1.1 Introduction
    1.1.1 Historical Perspective of ESI-MS for Measuring Protein Complexes
    1.1.2 Types of Interactions that Are Probed by ESI-MS
1.2 Critical Aspects of the Experimental Procedure
    1.2.1 Instrumental Parameters
        1.2.1.1 Electrospray Ionization Source
        1.2.1.2 Atmosphere/Vacuum Interface and Pressure
        1.2.1.3 Mass Spectrometry Analyzers
        1.2.1.4 Ion Mobility Analyzers
    1.2.2 Sample Preparation
1.3 Solution Phase Equilibria and Gas Phase Dissociation
    1.3.1 Measuring Solution Dissociation Constants
    1.3.2 Tandem Mass Spectrometry of Protein Complexes

1.4 Conc lusions

   Acknowledgments

   References

1.1 INTRODUCTION

Beyond its primary, secondary, and tertiary structures, the quaternary struc-ture of a protein can be defined as its interactions and associations with other proteins, macromolecules, a nd ligands that conspire to de fine its biological function. Thus, the structural determination of protein complexes can play an important role in the fundamental understanding of biochemical pathways. Traditionally, researchers have a variety of tools at their disposal to probe and measure such interactions. These tools include ultracentrifugation, light scat-tering, yeast two-hybrid, surface plasmon resonance, affinity chromatogra-phy, and native gel electrophoresis, and the methods that provide an "image" of the protein complex, such as cryoelectron microscopy, nuclear magnetic resonance (NMR) spectroscopy, and X-ray crystallography. Eac h of these methods has its advantages and disadvantages, and each provides a defined level of information de tail, from low-resolution assembly size information (e.g., dynamic light scattering) to h igh-resolution structure from NMR and X-ray.

Mass spectrometry (MS) is becoming a tool for probing noncovalently bound protein–ligand associations. I ts popularity i s i ncreasing for several reasons, i ncluding t he i mpressive res ults f rom a n umber o f resea rchers worldwide, including Carol Robinson [1] and Albert Heck [2], who have demonstrated the capabilities of MS to measure protein complexes as large as the 2 MDa ribosome [3]. In addition, the general field of proteomics has featured prominently and has encouraged more biochemical scientists to ap-ply mass spectrometry into their research strategies. Perhaps the greatest in-centive for the increasing interest in mass spectrometry is the improvements in the technology; sensitivity, resolving power, and mass accuracy have been improving steadily, and the availability of more MS systems tailored to spe-cific requirements (e.g., laboratory space, budget) is increased. Although most of the improvements have targeted peptide-centric analysis for protein sequencing and identification, these improved features have benefited also the analysis of intact proteins and protein complexes.

As demonstrated by the pioneering work of John Fenn, who was awarded the Nobel Prize in Chemistry in 2002 for his development of electrospray ionization (ESI) [4], taking liquid solutions and aerosolizing them into the vapor state has unique advantages for measuring large biomolecules. Not only can the molecular weight of proteins be measured very accurately, especially with higher resolution mass spectrometers, such as the time-of-flight (TOF) analyzer, the Fourier transform ion cyclotron resonance (FT-ICR) mass spectrometer and, more recently, the OrbiTrap analyzer, but sequence information can be derived, either from the intact protein directly (e.g., top–down sequencing) or from proteolytic fragments (e.g., bottom–up sequencing) in combination with tandem mass spectrometry (MS/MS). However, the solution phase origin of ESI-MS is a unique advantage, compared to matrix-assisted laser desorption/ionization (MALDI), for the analysis of protein complexes. Most protein interactions important to biology persist in an aqueous environment under so-called "physiological" conditions. The majority of biophysical methods used to probe protein complexes in vitro cannot accommodate all of the biochemicals necessary to define the "physiological" state of a cell. (As an example, the next time you run across a paper describing the high-resolution X-ray crystal structure of a protein, read the conditions necessary to crystallize the protein. When was the last time one encountered polyethylene glycol in a cell?) It is assumed that the structure of many proteins in a water environment and at near neutral pH is not perturbed significantly compared to their physiological state. This allows ESI-MS to analyze directly proteins in aqueous solution at near neutral pH. In some examples, the secondary and tertiary structures can be probed by gas phase methods, such as hydrogen–deuterium exchange and ion mobility. Moreover, the protein interactions are sufficiently retained upon the transition to the gas phase that the size and binding stoichiometry can be measured. Thus, the ability of ESI to ionize macromolecules without disrupting covalent bonds and maintaining the weak noncovalent interactions is a key distinguishing feature of ESI for the study of biological complexes [5]. The molecular mass measurement provides a direct determination of the stoichiometry of the binding partners in the complex, even for multiligand heterocomplexes.

1.1.1   Historical Perspective of ESI-MS for Measuring Protein Complexes

Peptide and protein associations have been reported throughout the literature of biological mass spectrometry. From the early days of field desorption/ionization (FD and FI), fast atom bombardment (FAB or liquid secondary ionization mass spectrometry, LSIMS), particle beam and thermospray, electrohydrodynamic-based desorption/ionization (EHD), laser desorption, and californium-252 plasma desorption, curious "adducts" have been observed in the mass spectra of peptides and proteins. In many cases, adducts, or the apparent binding of another atomic or molecular species, were associated with trace levels of alkali or alkali-earth salts, such as sodium, potassium, lithium, and calcium. The binding of ubiquitously present salts helped promote the formation of ionized peptides and proteins for their observation by mass spectrometry. Also present in the mass spectra in some cases were peaks that were assigned as a peptide "dimer," such as $(2M + H)^+$ or $(2M + Na)^+$. Such observations were explained usually as a result of nonspecific aggregation in the gas phase. The local analyte concentrations in the desorption/ionization region of the MS source were sufficiently high to promote the formation of random associations.

Such chance associations were observed also in the early days of ESI. Myoglobin and 12 kDa cytochrome c were (and still are) common test proteins for ESI-MS, primarily because of their relative high purity from commercial sources and their economical prices. Myoglobin is a 153 amino acid polypeptide chain that functions as an oxygen carrier through its noncovalent association with a heme (protoporphyrin IX) molecule. Cytochrome c similarly binds heme, but through covalent thioester bonds. Using denaturing solution conditions to perturb noncovalent heme–protein associations, such as 50% acetonitrile or methanol and high acid concentrations (pH 3 or lower), ESI mass spectra of myoglobin show multiply charged molecules for the apoprotein, whereas cytochrome c retains heme binding because of its covalent association. However, sometimes a set of low abundant peaks representing the binding of heme to myoglobin were observed. In addition, peaks for protein and peptide dimers can be observed, particularly if the analyte concentrations were relatively high in solution (ca. 25 μM or higher).

The first report of specific associations probed by ESI-MS was authored by the Cornell University groups of Ganem and Henion. The intact receptor–ligand complex between FK binding protein (FKBP) and macrolides rapamycin and FK506 [6], and the enzyme–substrate pairing between lysozyme and N-acetylglycosamine (NAG) and its cleavage products were reported [7]. Several reports of other biochemical noncovalently bound systems using ESI-MS detection followed shortly afterwards, including the ternary complex between the human immunodeficiency virus (HIV) protease dimer protein binding to a substrate-based inhibitor [8]. The ESI-MS of the noncovalent heme–myoglobin complex was reported first by Katta and Chait [9]; dramatic differences in the myoglobin spectra measured from aqueous solutions between pH 3.35 and pH 3.90 were observed. Myoglobin is fully denatured at pH 3.35, and the mass spectrum shows only ions for the apo or nonbinding form of the protein. The protein exists in its native configuration at pH 3.90, thus allowing the protein to fold properly, and the noncovalent binding of a heme molecule occurs. The effect of solution thermodynamics and the relative stability of the gas phase complex was studied shortly afterwards [10, 11]. Ribonuclease S (RNase S) is composed of the hydrophobically bound 11.5 kDa S-protein complexed to 2 kDa S-peptide. In solution, S-peptide binds to S-protein with a solution binding constant, K_D, around 1–10 nM; the solution temperature dependence on the gas phase stability was predicted from thermodynamic parameters.

These early examples established not only the feasibility of the ESI-MS method, but also the design of the experiment to ensure validation of the observations. The validity of the results needs to be established in order to assess a meaningful interpretation and link to the solution phase system. These papers also helped move biology, biochemistry, and medicinal chemistry to the forefront of applications for ESI-MS technology.

As smaller protein–ligand complexes were found to be amenable to ESI-MS measurements, the ability to access larger molecular weight complexes was tested. However, as larger complexes were tested, it was found that the relative charging is low compared to the accessible mass-to-charge (m/z) range of instruments employed by most laboratories during these early days of ESI. For example, shown in Figure 1.1 is the ESI-MS of yeast alcohol dehydrogenase, a homo-tetrameric protein complex of 147 kDa; molecular ions are observed above m/z 4000.

The quadrupole mass analyzer was the system favored by John Fenn as ESI was being developed. One of the advantageous features of ESI is the multiple charging that allows most mass analyzers of limited m/z range to be used for biomolecule analysis. Most quadrupole mass analyzers are of limited

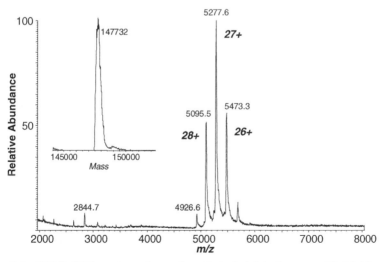

Figure 1.1   ESI QqTOF mass spectrum of yeast alcohol dehydrogenase (5 µM, 10 ammonium acetate, pH 6.5). The inset shows the mass deconvoluted spectrum, indicating a molecular weight of 147.7 kDa for the intact tetrameric complex.

m/z range, typically less than 4000. However, multiply charged ions for complexes such as protein–protein quaternary complexes exhibit relatively low charge at high m/z. The amount of charging that a biomolecule exhibits in an ESI mass spectrum has been correlated to a global solution structure [12]. The relatively narrow charge distribution of a low-charge state (typically four to five charge states) represents retention of the higher order structure of the native protein complex, presumably because fewer charge sites are exposed and/or the Coulombic restraints restrict charging for the compact structure. A magnetic sector ESI mass spectrometer, in general, has sufficient m/z range (to m/z 10,000) to study protein complexes such as alcohol dehydrogenase and pyruvate kinase and other quaternary protein structures [13, 14]. However, the sensitivity and resolution at very high m/z of the time-of-flight (TOF) analyzer provides an ideal system for large noncovalent complexes. This was first demonstrated by Standing and co-workers with the large protein complexes from soybean agglutinin [15] and extended by a variety of other protein systems [16]. Today, the ESI-TOF and the quadrupole time-of-flight (QTOF) mass spectrometers are the systems of choice for most ESI measurements of protein noncovalent complexes [17, 18].

1.1.2 Types of Interactions that Are Probed by ESI-MS

The fundamental forces of almost all noncovalent interactions in water include hydrophobic effects, hydrogen bonds, salt bridges, van der Waals interactions, and Coulombic interactions. Yet, these types of interactions that govern noncovalent binding in solution sometimes play only a limited role in the observed MS results from the ESI-MS gas phase measurements. The transition from a high-dielectric environment (i.e., water) to a solventless vacuum environment strengthens electrostatic interactions, and thus complexes held together by electrostatic interactions are extremely stable in the gas phase. Protein–nucleic acid complexes, noncovalent complexes between a highly positively charged molecule and a negatively charged macromolecule such as human immunodeficiency virus (HIV) Tat peptide–TAR RNA complex and the NCp7 – ψ-RNA complex, are extremely stable, as the complexes are not observed to dissociate at very high collision energies [19–21]. Hydrophobic interactions in solution appear to be weakened in vacuum. For example, the relative affinities measured by ESI-MS for small molecule hydrophobic binding to acyl CoA-binding protein do not correlate with their solution affinities [22]. The differences between electrostatic and hydrophobic interactions in the gas phase are further highlighted by inhibitor binding studies to HIV-1 TAR RNA [23]. Positively charged aminoglycosides such as neomycin are known to bind to RNAs through charge–charge interactions. The neomycin–TAR RNA complex was not observed to dissociate in the gas phase. However, inhibitors with similar solution binding affinities to TAR RNA that bind through hydrophobic-type means are extremely labile in the gas phase.

The strengthened role of charge–charge electrostatic forces in gas phase stabilities can be exploited for measuring weak, solution phase interactions. The pathological hallmark of the neurodegenerative disorder Parkinson's disease (PD) is the presence of intracellular inclusions, called Lewy bodies, in the dopaminergic neurons of the substantia nigra. Filamentous α-synuclein (AS, M_r 14460) protein is the major component of these deposits and its aggregation is believed to play an important role in Parkinson's disease. AS binds to natural polycations, such as spermidine and spermine. A previous NMR study suggested that spermine (M_r 202) binds to the C-terminal acidic region of AS with a solution binding affinity (K_D) of 0.6 mM [24]. ESI-MS with a QqTOF system shows the ability to measure binding affinities in the low millimolar range for the 1:1 AS–spermine complex [25]. The stability for such weakly bound ligands is enhanced in the gas phase because of charge–charge electrostatic interactions.

Likewise, protein–metal ion binding can be quite stable in the gas phase. For example, human superoxide dismutase (SOD) is a small 32 kDa homodimeric protein that binds transition metal ions, such as zinc and manganese. Each protein monomer has two metal binding sites. Figure 1.2 shows the ESI mass spectrum of human SOD in the presence of excess zinc; multiply charged SOD dimer proteins are observed to bind to four zinc metals

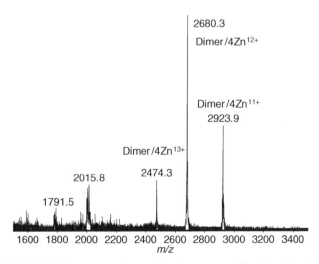

Figure 1.2 ESI QqTOF mass spectrum of human superoxide dismutase (SOD; 10 mM ammonium acetate, pH 7.5) in the presence of excess zinc chloride. Multiply charged ions for the SOD dimer bound to four zinc ions are labeled.

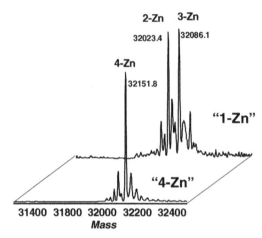

Figure 1.3   ESI QqTOF mass deconvoluted spectra of human superoxide dismutase (SOD; 10 mM ammonium acetate, pH 7.5) in the presence of excess zinc chloride (labeled as "4-Zn") and substoichiometric zinc (labeled as "1-Zn").

in total. One can titrate zinc metal into a solution of apo-SOD dimer to observe metal binding occupancy (Figure 1.3). Dissociating the gas phase SOD dimer–metal ion complex results in protein monomer release while retaining two zinc metal ions per monomer; that is, the protein–metal interaction is stronger than the protein–protein interaction in the gas phase.

The different relative stabilities of gas phase interactions have implications for using ESI-MS to determine solution phase absolute and relative binding affinities. For compounds that bind to a target molecule with similar type binding mechanisms, and thus may have similar gas phase stabilities, determining their relative binding affinities by ESI-MS should not be problematic. However, if hydrophobic interactions are in play, the lability of the gas phase complex may conspire to reduce the confidence of the MS data.

1.2   CRITICAL ASPECTS OF THE EXPERIMENTAL PROCEDURE

1.2.1   Instrumental Parameters

For the most part, the researchers active in the early days of ESI who explored its application for measurement of noncovalent protein complexes found the experimental parameters critical for the success of the experiments. Besides new developments in MS analyzers that offer higher sensitivity, resolution, and m/z range for larger protein complexes, the critical parameters found during these early days persist in today's experiments.

1.2.1.1   Electrospray Ionization Source So lution flow rates of 1–10 μL/min were common for a majority of ESI-MS applications, and this was used also for the measurement of protein–ligand complexes. However, difficulties for ESI of aqueous solutions were found until the development of nanoelectrospray. Work by Caprioli [26] and Mann [27] demonstrated the effectiveness of ESI at significantly lower flows, down to 1 0–200 nL/min, and this played a significant role in not only ESI-MS of biomolecules in general, but also in the study of noncovalent complexes. The advantages of nanoliter per minute analyte flow includes not only reduction of the overall consumption of precious sample without compromising signal intensity, but also the generation of smaller droplets, which results in increased signal levels.

Nanoelectrospray helps in the requirement for droplet desolvation for noncovalent complex studies. Desolvation of ESI from aqueous solutions is not as easy as found for aqueous/organic solvent mixtures because of reduced volatility. The generation of smaller diameter droplets from small-orifice ESI needles aids the desolvation process. For some examples, adding heat externally to the ESI spray region (e.g., heating of a countercurrent gas flow) may also help desolvate the aqueous droplets. However, depending on the solution and gas phase stabilities of the complex, increasing solvent/droplet temperatures may destabilize the noncovalent complex. In fact, examples have been reported that utilize solution cooling to improve the stability of the complex to be measured [28].

1.2.1.2   Atmosphere/Vacuum Interface and Pressure   Nearly all types of atmospheric pressure/vacuum interfaces for ESI-MS have been used successfully for the analysis of noncovalent protein complexes. These range from nozzle–skimmer interfaces, heated metal or glass capillary inlets, to the orthogonal "Z-spray" interface used on Micromass/Waters systems. For all cases, optimal tuning of each of the various parameters associated with each interface type is critical for efficient transmission of the noncovalent protein ions. It is somewhat analogous to a restaurant waiter balancing an egg on his/her head while walking quickly between tables. The waiter could run to each table, but it does no good if the egg drops to the floor, breaking open its contents. Each lens within the interface has its optimal settings for transmitting the highest intensity ion beam, but each region between lenses can subject the fragile protein complex to collisional dissociation [13, 29]. Moreover, ion desolvation is effected in the interface region. Maximal desolvation to generate the narrowest spectral peaks is desired, while minimizing apparent dissociation of the noncovalent complex. Because of small differences in the geometry and vacuum pressures of each ESI interface, tuning conditions between instruments of the same interface may differ slightly.

Counter to traditional mass spectrometry philosophies that encourage high vacuum for establishing high performance, ESI-MS of large proteins utilizes low vacuum in the interface region while somehow utilizing high vacuum in the measurement/detector region, which requires nine orders of magnitude in differential pumping. Perhaps it is fitting for a gas phase technique that measures analytes originating from solution, but several reports have shown that transmission of high-mass ions requires pressures in the first vacuum stages of the mass spectrometer to be increased by reducing the pumping speed or by adding a collision gas in the collision quadrupole of the QTOF. Krutchinsky et al. [30] have suggested that larger ions may acquire substantial kinetic energies (of more than 1 keV) when they are electrosprayed out of the supersonic jet. This may have a negative effect on the transfer of ions and the orthogonal extraction into the TOF region. The increased pressure in the preceding quadrupoles/hexapoles may act as a collisional dampening interface. The enhancement is most noticeable for very large assemblies observed at high m/z, as demonstrated by the reports on protein complexes in excess of 1 MDa [18, 31]. It is now widely accepted that a combination of collisional dampening, increased cooling of the ions, and more efficient desolvation is critical for the sensitive detection of large ions at high m/z [32].

1.2.1.3 Mass Spectrometry Analyzers   Although the majority of current ESI-MS research projects for studying noncovalent protein complexes utilize time-of-flight (or quadrupole TOF) analyzers [33], there is no inherent operational characteristic of the analyzer that limits its use for such studies. However, the accessible m/z range (and its associated sensitivity and resolution) is the overriding factor when choosing the appropriate system. Nearly all types of mass analyzers have been used for these types of studies. These range from single and triple quadrupoles [10], forward- and reverse-geometry magnetic sector analyzers [13], quadrupole and linear ion traps [34], TOF and QTOF [35], to Fourier transform ion cyclotron resonance (FT-ICR) instruments [36]. The ion measurement timescale ranges from microseconds to milliseconds, with no apparent correlation between timescale and performance. There are distinct advantages for using analyzers with tandem mass spectrometry (MS/MS) capabilities, as dissociation of the gas phase complex can yield information on the nature of the ligand association (see later discussion).

In general, however, the m/z range and the overall sensitivity of the analyzer are the overarching factors when selecting an appropriate instrument. In general, the larger the molecular weight of the protein complex, the larger the m/z range needed to measure the full envelope of multiply charged molecules. This is especially true for native proteins and protein complexes as discussed earlier. Native protein mass spectra show typically only a few charge states and much reduced absolute charging compared to their denatured forms. Most denatured protein mass spectra show multiple charging in the 800–3000 m/z range, regardless of their size. Thus, even for a relatively small protein–ligand complex, s uch a s t he 1 7.5 kDa m yoglobin–heme co mplex, t he 8 +-charged molecule would be observed around m/z 2200, outside of many quadrupole and ion trap analyzers (although the 9+ and 10+ molecules would be within the available m/z 2000 range) [37]. Because of the efficient transmission of higher m/z ions, TOF and QTOF analyzers are the popular choices for measuring l arger pro tein co mplexes. S tandard TO F a nd QTO F a nalyzers ha ve m/z ranges of 5000–10,000 that are well within the range necessary for many protein complexes. However, for very large protein complexes, such as larger than 0.5 MDa, m/z ranges above 10,000 may be necessary. Kaltashov has empirically found a nea r l inear re lationship between $\ln(N)$ versus $\ln(S)$, where N is average charge for a native protein and S is surface area based on available cr ystal s tructures [12]. T hus, for a 6 90 kDa 20S pro teasome co mplex, composed of 14 α-subunits and 14 β-subunits, multiple charging for the fully assembled 28 -mer is obser ved a round m/z 11,000 with an average of 63+ charges [35]. Dissociation of the 28-mer generates ions for the 27- and 26-mer aggregates that span from m/z 15,000 to 35,000 (Figure 1.4).

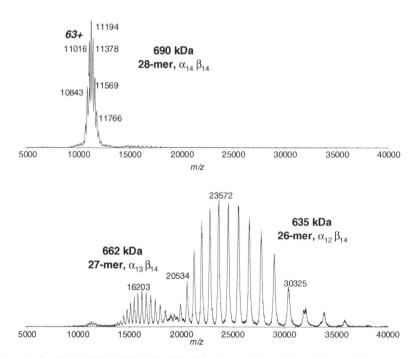

Figure 1.4 (Top) ESI-QqTOF-MS of the *M. thermophila* $\alpha_7\beta_7\beta_7\alpha_7$ 28-mer 20S proteasome with an orifice potential of +100 V. (Bottom) ESI-QqTOF-MS of the 20S proteasome with the orifice potential of +300 V. Dissociation in the atmosphere/vacuum interface of the $\alpha_7\beta_7\beta_7\alpha_7$ complex y ields the liberated 27.4 kDa α-subunit (not shown) and the remaining $\alpha_7\beta_7\beta_7\alpha_6$ (or $\alpha_{13}\beta_{14}$) and $\alpha_{12}\beta_{14}$ complexes.

Although FT-ICR analyzers have demonstrated impressive resolution capabilities, well over 1 million resolving power, very few studies have demonstrated comparable high-resolution results for noncovalent protein complexes above the m/z 2000 limit. Similarly, the newer OrbiTrap analyzer is capable of resolution above 100,000 with low parts-per-million mass accuracy [38]. But to date it has not demonstrated comparable performance for measurement of ions above m/z 3000. Thus, it remains to be seen whether high-resolution FT-ICRs and OrbiTraps will be applied for studies of larger protein complexes. However, for small proteins (e.g., 20–35 kDa) binding to smaller sized ligands (e.g., products from combinatorial chemistry libraries), these analyzers may be of special utility for ligand screening [36].

1.2.1.4 Ion Mobility Analyzers   Ion mobility spectrometry (IMS) is an electrophoretic technique that allows ionized analyte molecules to be separated on the basis of their mobilities in the gas phase, as opposed to separation based on their mass-to-charge ratio in conventional mass spectrometry. However, coupling IMS with MS forms a powerful combination for examining protein conformers and, potentially, protein complexes. For example, Clemmer and Cooks have combined a desorption electrospray ionization (DESI) source to an ion mobility time-of-flight mass spectrometer for the analysis of proteins [39]. Analysis of 12 kDa cytochrome c and 14 kDa lysozyme proteins with different DESI solvents and conditions shows similar mass spectra and charge state distributions to those formed when using electrospray to analyze these proteins in solution. The ion mobility data show evidence for compact ion structures (when the surface is exposed to a spray that favors retention of native-like structures (50:50 water:methanol)) or elongated structures (when the surface is exposed to a spray that favors "denatured" structures (49:49:2 water:methanol:acetic acid)).

Similarly, Bowers and Gray studied the protein α-synuclein, implicated in Parkinson's disease, ESI-MS, and ion mobility [40]. It was found that both the charge-state distribution in the mass spectra and the average protein shape deduced from ion mobility data depend on the pH of the spray solution. Negative ion ESI-MS of pH 7 solutions yielded a broad charge-state distribution centered at 11−, and the ion mobility data is consistent with an extended protein structure. Data obtained for pH 2.5 solutions, on the other hand, showed a narrow charge-state distribution centered at 8−, and ion mobilities in agreement with compact α-synuclein structures. The average cross section of α-synuclein at pH 2.5 is 33% smaller than for the extended protein sprayed from pH 7 solution. Significant dimer formation was observed when sprayed from pH 7 solution but no dimers were observed from the low-pH solution.

ESI-IMS, however, has not been applied as extensively to the measurement of noncovalent complexes compared to mass spectrometric detection.

Colgrave et al. [41] reported on the noncovalent complexes formed between cyclic 12-crown-4, 15-crown-5, and 18-crown-6, and acyclic polyethers with amino acids (histidine and arginine) and peptides (MRFA, MFAR, and bradykinin) using (nano-ESI) ion mobility spectrometry. The reduced mobilities for these complexes were observed and correlate well with the mass and size of the polyether. They demonstrated the ability of IMS to distinguish between cyclic and acyclic polyethers and their complexes with biomolecules based on differences in their reduced mobilities. These differences are attributed to variations in the collision cross section arising from subtle changes in conformation in these ligand–receptor complexes.

Larger protein complexes have been analyzed by an ion mobility device, termed Gas-Phase Electrophoretic Mobility Molecular Analyzer (GEMMA) [42]. The GEMMA utilizes a differential mobility analyzer (DMA) to measure gas phase electrophoretic mobility (EM) that is proportional to the electrophoretic diameter of the particle in air. The GEMMA offers utility for the characterization of proteins, glycoproteins, protein aggregates, high-mass noncovalent protein complexes, whole viruses, and nanoparticles of biological importance [42]. For GEMMA, the biomolecules are electrosprayed followed by charge neutralization of the evaporating droplets to generate primarily neutral and singly charged molecules. Alpha-particles generated by a ^{210}Po reactor ionize gas molecules in the atmosphere, producing reactive species such as H^+, H_3O^+, and $(H_2O)_nH_3O^+$. These primary species quickly form ionized clusters 1–2 nm in size, chiefly with water molecules in the atmosphere. The clusters diffuse to the evaporating droplets, causing their charge distribution to approach a distribution centered about zero charge. When the droplets have evaporated completely, the distribution consists almost entirely of neutral macromolecules and singly charged macroions. The singly charged protein molecules are size separated through a scanning DMA according to their EMs in air, and are detected by a condensation particle counter (CPC). Their mobilities are interpreted in terms of an "electrophoretic mobility diameter" (EMD) of the gas phase protein.

The electrophoretic mobility of a particle is governed by its size and shape, and this method has been used also to characterize proteins and noncovalently bound protein complexes, showing a correlation between the experimentally derived electrophoretic mobility diameter and its predicted molecular mass [42]. The resolving power of GEMMA is approximately 10–20 in terms of the EM diameter, but this does not preclude the utility of the GEMMA measurement for large proteins. Mass measurements are based on a simple model relating molecular weight to the diameter of a sphere and an effective density. From the GEMMA measurements by our laboratory and from those reported by Bacher et al. [42] for over 50 protein complexes ranging in size from small

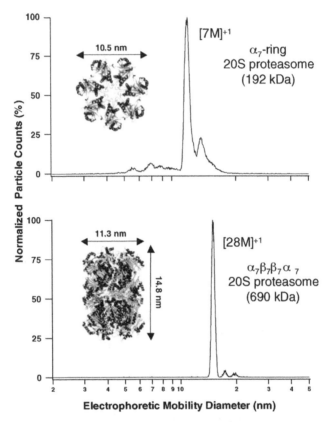

Figure 1.5   ESI-GEMMA of the α_7 and $\alpha_7\beta_7\beta_7\alpha_7$ 20S proteasome complexes from *M. thermophila*. The protein structures show the dimensions as measured by X-ray crystallography.

protein d imers to co mplexes a s l arge a s t he 6 90 kDa 20S pro teasome a nd MDa-range vi ral pa rticles, a n e ffective de nsity o f a pproximately 0 .6 g/cm^3 can be used to estimate the molecular masses of proteins.

For e xample, E SI-GEMMA ha s b een u tilized to de tect d ifferences i n g as phase electrophoretic mobility between the empty 20S pr oteasome and the 20S proteasome encapsulating protein substrates. The gas phase EMDs of the cylindrical proteasome and the disk shaped α-ring from *Methanosarcina thermophila* were consistent with crystal structure determined dimensions of the complexes from o ther a rchaea (Figure 1 .5) [35]. To "trap" t he s ubstrates wi thin the pro-teasome, the proteolytic activity of the complex was irreversibly inhibited prior to incubation of the complex with the protein substrates. Based on the change in GEMMA-determined molecular weight, an average of 4.5 substrate proteins were estimated to be sequestered within the complex [35]. High-resolution MS data has since shown that the 20S proteasome from *Thermoplasma acidophilum* can sequester a maximum of three or four substrate proteins of similar size [43].

1.2.2   Sample Preparation

The most critical points for the ESI-MS analysis of noncovalent protein complexes a re ma intaining pro per s olution co nditions f or k eeping t he pro tein complex in its folded and functional native state and effective desolvation of the E SI-generated d roplets. U sing the pro per a queous s olvents, p H (almost exclusively near neutral physiological pH, but for some acidic proteins such as HIV protease, "native" folding and activity is maintained at acidic pH [20]), and ionic strength buffer systems are necessary to ma intain complexation. Deviation from optimal solution conditions may reduce the observed relative proportion of complex formation. However, in some cases, it may be necessary to compromise solution conditions optimal for protein activity for the MS analysis because many buffers are not compatible with ESI. Volatile buffers such as ammonium acetate and ammonium bicarbonate are the most popular choices for such ESI-MS experiments because they do not often form extensive gas phase adducts with the macromolecules (as do phosphate- and sulfate-based buffers) and background ion formation is reduced without significant re duction i n protein i on formation. B uffer co ncentrations a re t ypically at t he 5–200 mM concentration levels, but exceptions may occur, such as some protein–DNA complexes. Best sensitivity for detecting the noncovalent protein complex is obtained using nanoelectrospray sources with borosilicate or glass nanospray needles (50–200 nL/min) because of the smaller droplets formed.

Most critical to the success of the analysis is the purity and quality of the protein sample. Compared to t he E SI-MS s ignal l evels mea sured for p ure proteins under common denaturing conditions (e.g., 50% by volume of acetonitrile or methanol with low concentrations of acetic acid or formic acid), the signals measured for proteins and their complexes ca n be reduced by a factor of 10 or more using "native" solution conditions, such as pH 6–8 aqueous a mmonium ace tate s olutions. O ther m olecular e ntities t hat ca n e ffectively compete for the available droplet surface charges, such as peptide and protein contaminants and other small molecules, will further serve to reduce signal i ntensities. Ex tensive d ialysis a nd t he use o f ce ntrifugal me mbrane filtration a re po pular me thods f or s alt re moval a nd s ample co ncentrators. All common detergents (cationic, anionic, and zwitterionic) are not tolerated well for these ESI (MS and ion mobility) experiments. Adducts formed by binding of salts (e.g., cationic sodium and potassium, and anionic phosphate and sulfate) further reduce the overall sensitivity by spreading the signal for the protein over many more channels than for the multiply protonated forms. For very large complexes in which salt adducts a re not fully resolvable by mass spectrometry, adduct formation increases peak widths and can shift the peaks to h igher m/z values; this re duces t he ab ility to mea sure acc urately the molecular weight of the protein and protein complex. A recent strategy reported by Robinson and colleagues may help reduce some of the problems associated with adduct formation [3]. By measuring adduct formation for a variety of large protein complexes, a simulation and modeling method is developed to describe and interpret the electrospray mass spectra of large noncovalent protein complexes. Using this method, the mass accuracy for large protein complexes up to the 2 MDa ribosome is significantly improved.

1.3   SOLUTION PHASE EQUILIBRIA AND GAS PHASE DISSOCIATION

1.3.1   Measuring Solution Dissociation Constants

The correlation between the ESI-MS gas phase measurements and the solution phase characteristics has extended the application of ESI-MS to the determination of solution relative and absolute equilibrium binding constants. Competitive binding experiments, in which the total ligand concentration (single or a mixture of ligands) is greater than or equal to the protein receptor, can measure the relative binding affinities for a mixture of ligands [44]. Absolute binding constants can be derived by titration experiments and by construction of Scatchard binding plots. This was first demonstrated by Henion's group for measuring the binding constants of vancomycin antibiotics with peptide ligands [45]. The equilibrium dissociation constants of the 96 kDa dimer and 287 kDa hexameric oligomeric forms of citrate synthase binding to NADH, an allosteric inhibitor of the enzyme, was determined by Duckworth and coworkers using an ESI-TOF instrument [46]. Griffey measured the dissociation constants for oligonucleotide binding to albumin [47]. Similarly, they demonstrated the applicability of high-resolution FT-ICR mass spectrometry for measuring small molecule binding to RNA targets, and they use this strategy to screen small molecule inhibitors [48]. With ESI sources that generate stable and reproducible ion currents between multiple samples, the process of titration experiments for the purpose of measuring binding constants can be automated [49].

1.3.2   Tandem Mass Spectrometry of Protein Complexes

In general, collisionally activated dissociation (CAD) of multiply charged protein complexes held together through noncovalent bonding yields the liberated protein(s) and ligand(s). This is understandable because these are the weakest bonds found throughout the complex in solution. Also, although the results of this type of experiment may be rather "uninteresting," it can be quite useful analytically for unknown protein–ligand systems. For example, given a solution of a single unknown protein that contains also a potential smaller molecule ligand, ESI-MS of the protein under denaturing solution conditions (e.g., 50% acetonitrile with 2% acetic acid) yields information on the molecular weight of the denatured protein. The presence of the liberated small molecule ligand may be in doubt because it is likely that many peaks appear in the m/z 100–800 range that may represent the unknown ligand. The follow-up experiment would be to acquire an ESI mass spectrum of the solution under native conditions (e.g., 10 mM ammonium acetate, pH 6.8) and measure the mass of the protein–ligand complex. This can be followed by an MS/MS experiment, in which the precursor ion for the protein–ligand complex is selectively dissociated by CAD or perhaps infrared multiphoton dissociation (IRMPD) if performed with an FT-ICR analyzer. The combination of these experiments should yield the mass of the putative binding ligand, even if multiple ligands are bound simultaneously. This is an effective experiment for mixtures of putative ligands, for example, from combinatorial libraries, as demonstrated for carbonic anhydrase and an SH2 domain protein by Smith and co-workers [50] and by Marshall and co-workers [51], respectively.

For large multiprotein complexes, MS/MS generates a liberated monomer (or a few subunits) and the remaining, much larger complex (minus the liberated molecule). For example, Figure 1.6 shows the ESI mass spectrum of the 52 kDa homo-tetrameric streptavidin complex. Although most tetrameric protein complexes are believe to be composed of a dimer of dimer proteins in solution, such as that found for streptavidin, tandem mass spectrometry of the streptavidin tetramer liberates the monomer, leaving behind a trimer

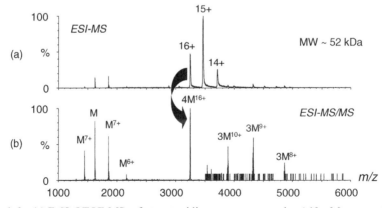

Figure 1.6 (a) ESI-QTOF-MS of streptavidin tetramer complex (10 mM ammonium acetate). (b) ESI-MS/MS of the 16+ tetramer protein, yielding product ions for the released monomer and the remaining trimer protein.

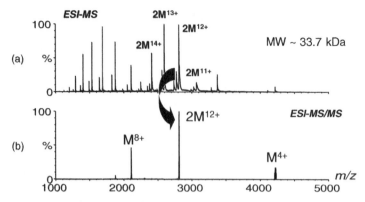

Figure 1.7   (a) E SI m ass sp ectrum o f t he i nterferon-γ p rotein d imer (10 mM a mmonium acetate). (b) ESI-MS/MS of the 12+-charged dimer protein y ields the 4+- and 8+-charged monomer products.

gas phase complex. Thus, the characteristics of the gas phase assembly may not match that found in solution. On the other hand, the dissociation mass spectrum of the 20S proteasome shown in Figure 1.4 is somewhat consistent with the general topology of the complex. Based on a $\alpha_7\beta_7\beta_7\alpha_7$ geometry, the loss of the outer α-subunits would be favored more compared to loss of the internal β-subunits because of the reduced number of potential intermolecular protein–protein contacts.

Furthermore, the distribution of charge in the products may not be evenly distributed. Jurchen and Williams [52] have reported that the asymmetric charge distribution results from unfolding of the monomer product, thus exhibiting a more flexible conformation [52]. This is demonstrate in Figure 1.7 for the MS/MS dissociation of the 34 kDa interferon-γ homodimer. CAD of the 12+ precursor m olecule y ields 8+ and 4+-charged m onomer products, rather than two 6+ products. Based on current hypotheses, the 8+ monomer is released as an unfolded, or more denatured, product. However, whether this type of experiment can yield meaningful information regarding the solution structure of the protein complex remains to be proved.

1.4   CONCLUSIONS

Mass spectrometry-based methods have the potential to provide a better understanding of the relationship between the structure of protein complexes and their biological function. Francis Collins of the National Human Genome Research Institutes states that "genes and gene products do not function independently, but participate in complex, interconnected pathways, networks and molecular systems that, taken together, give rise to the workings of cells, tissues, organs and organisms. Defining these systems and determining their properties and interactions is crucial to understanding how biological systems function" [53]. The systematic identification and characterization of these "machines of life" will "provide the essential knowledge base and set the stage for linking proteome dynamics and architecture to cellular and organismic function" [54]. Tools based on measurement of the gas phase macromolecule will be complementary to large-scale efforts in structural biology to determine the structure of all biologically important proteins and complexes. As improvements to all aspects of the experiment, from brighter ionization sources to more sensitive and higher resolution analyzers, are made available to scientists, the ability of mass spectrometry to directly impact biomedical research will improve.

ACKNOWLEDGMENTS

We are grateful for the support of the UCLA Functional Proteomics Center provided by the W. M. Keck Foundation, the funding of our research by the National Institutes of Health (RR20004 to JAL), and the U.S. Department of Energy for funding of the UCLA-DOE Institute for Genomics and Proteomics (DE-FC03–87ER60615). CSK was supported by the UCLA-NIH Chemistry–Biology Interface Training grant. We acknowledge research contributions from Shirley Lomeli and Sheng Yin (UCLA).

REFERENCES

1. Benesch, J. L. P., and Robinson, C. V. (2006). Mass spectrometry of macromolecular assemblies: preservation and dissociation. *Curr. Opin. Struct. Biol.* **16**: 245–251.
2. Heck, A. J. R., and van den Heuvel, R. H. H. (2004). Investigation of intact protein complexes by mass spectrometry. *Mass Spectrom. Rev.* **23**: 368–389.
3. McKay, A. R., Ruotolo, B. T., Ilag, L. L., and Robinson, C. V. (2006). Mass measurements of increased accuracy resolve heterogeneous populations of intact ribosomes. *J. Am. Chem. Soc.* **128**: 11433–11442.
4. Fenn, J. B. (2003). Electrospray wings for molecular elephants (Nobel Lecture). *Angew. Chem. Int. Ed.* **42**: 3871–3894.
5. Loo, J. A. (1997). Studying noncovalent protein complexes by electrospray ionization mass spectrometry. *Mass Spectrom. Rev.* **16**: 1–23.
6. Ganem, B., Li, Y. T., and Henion, J. D. (1991). Detection of noncovalent receptor–ligand complexes by mass spectrometry. *J. Am. Chem. Soc.* **113**: 6294–6296.

7. Ganem, B., L i, Y. T., a nd Hen ion, J. D. (1991). Obser vation o f n oncovalent enzyme–substrate and enzyme–product complexes by ion-spray mass spectrometry. *J. Am. Chem. Soc.* **113**: 7818–7819.

8. Baca, M., and Kent, S. B. H. (1992). D irect obser vation of a t ernary complex between t he d imeric en zyme H IV-1 prot ease a nd a s ubstrate-based i nhibitor. *J. Am. Chem. Soc.* **114**: 3992–3993.

9. Katta, V., and Chait, B. T. (1991). Obser vation of the heme–globin complex in native myoglobin by electrospray-ionization mass spectrometry. *J. Am. Chem. Soc.* **113**: 8534–8535.

10. Ogorzalek Loo, R. R., Goodlett, D. R., Smith, R. D., and Loo, J. A. (1993). Observation of a noncovalent ribonuclease S-protein/S-peptide complex by electrospray ionization mass spectrometry. *J. Am. Chem. Soc.* **115**: 4391–4392.

11. Goodlett, D. R., Ogorzalek Loo, R. R., Loo, J. A., Wahl, J. H., Udseth, H. R., and Smith, R. D. (1994). A study of the thermal denaturation of ribonuclease S by electrospray ionization mass spectrometry. *J. Am. Soc. Mass Spectrom.* **5**: 614–622.

12. Kaltashov, I. A., and Mohimen, A. (2005). Estimates of protein surface areas in solution by electrospray ionization mass spectrometry. *Anal. Chem.* **77**: 5370–5379.

13. Loo, J. A., Ogorzalek Loo, R. R., and Andrews, P. C. (1993). Primary to quaternary protein structure determination with electrospray ionization and magnetic sector mass spectrometry. *Org. Mass Spectrom.* **28**: 1640–1649.

14. Loo, J. A. (1995). Obser vation of la rge s ubunit prot ein complexes by e lectrospray ionization mass spectrometry. *J. Mass Spectrom.* **30**: 180–183.

15. Tang, X.-J., Brewer, C. F., Saha, S., Chernushevich, I., Ens, W., and Standing, K. G . (1994). I nvestigation o f prot ein–protein n oncovalent i nteractions i n soybean agglutinin by e lectrospray i onization tim e-of-flight mass spectrometry. *Rapid Commun. Mass Spectrom.* **8**: 750–754.

16. Chernushevich, I. V., Ens, W., a nd S tanding, K . G. (1998). I n *New Methods for the Study of Biomolecular Complexes* (NATO ASI Series, Ser. C) (W. Ens, K. G. Standing, and I. V. Chernushevich, Eds.), Vol. 510, pp. 101–116, Kluwer, Dordrecht, The Netherlands.

17. Tahallah, N., P inkse, M., Ma ier, C. S ., a nd He ck, A. J . (2 001). T he e ffect o f the source pressure on the abundance of ions of noncovalent protein assemblies in an electrospray ionization orthogonal time-of-flight instrument. *Rapid Commun. Mass Spectrom.* **15**: 596–601.

18. Van Berkel, W. J. H., Van Den Heuvel, R. H. H., Versluis, C., and Heck, A. J. R. (2000). Detection of intact megadalton protein assemblies of vanillyl-alcohol oxidase by mass spectrometry. *Protein Sci.* **9**: 435–439.

19. Loo, J. A., Sannes-Lowery, K. A., Hu, P., Mack, D. P., and Mei, H.-Y. (1998). In *New Methods for the Study of Biomolecular Complexes* (W. Ens, K. G. Standing, and I. V. Chernushevich, Eds.), pp. 83–99, K luwer, Dordrecht, The Netherlands.

20. Loo, J. A., Holler, T. P., Foltin, S. K., McConnell, P., Banotai, C. A., Horne, N. M., Mueller, W. T., Stevenson, T. I., and Mack, D. P. (1998). Application of electrospray ionization mass spectrometry for studying human immunodeficiency virus protein complexes. *Proteins Struct. Funct. Genet.* **Suppl. 2**: 28–37.

21. Sannes-Lowery, K. A., Hu, P., Mack, D. P., Mei, H.-Y., and Loo, J. A. (1997). HIV-1 Tat peptide binding to TAR RNA by electrospray ionization mass spectrometry. *Anal. Chem.* **69**: 5130–5135.

22. Robinson, C. V., Chung, E. W., Kragelund, B. B., Knudsen, J., Aplin, R. T., Poulsen, F. M., and Dobson, C. M. (1996). Probing the nature of noncovalent interactions by mass spectrometry. A study of protein–CoA ligand binding and assembly. *J. Am. Chem. Soc.* **118**: 8646–8653.

23. Sannes-Lowery, K. A., Mei, H.-Y., and Loo, J. A. (1999). Studying aminoglycoside antibiotic binding to HIV-1 TAR RNA by electrospray ionization mass spectrometry. *Int. J. Mass Spectrom.* **193**: 115–122.

24. Fernández, C. O., Hoyer, W., Zweckstetter, M., Jares-Erijman, E. A., Subramaniam, V., Griesinger, C., and Jovin, T. M. (2004). NMR of α-synuclein–polyamine complexes elucidates the mechanism and kinetics of induced aggregation. *EMBO J.* **23**: 2039–2046.

25. Xie, Y., Zhang, J., Yin, S., and Loo, J. A. (2006). Top–down ESI-ECD-FT-ICR mass spectrometry localizes noncovalent protein–ligand binding sites. *J. Am. Chem. Soc.* **128**: 14432–14433.

26. Emmett, M. R., and Caprioli, R. M. (1994). Micro-electrospray mass spectrometry: ultra-high-sensitivity analysis of peptides and proteins. *J. Am. Soc. Mass Spectrom.* **5**: 605–613.

27. Wilm, M., and Mann, M. (1996). Analytical properties of the nanoelectrospray ion source. *Anal. Chem.* **68**: 1–8.

28. Benesch, J. L. P., Sobott, F., and Robinson, C. V. (2003). Thermal dissociation of multimeric protein complexes by using nanoelectrospray mass spectrometry. *Anal. Chem.* **75**: 2208–2214.

29. Smith, R. D., and Light-Wahl, K. J. (1993). The observation of non-covalent interactions in solution by electrospray ionization mass spectrometry: promise, pitfalls and prognosis. *Biol. Mass Spectrom.* **22**: 493–501.

30. Krutchinsky, A. N., Chernushevich, I. V., Spicer, V. L., Ens, W., and Standing, K. G. (1998). Collisional damping interface for an electrospray ionization time-of-flight mass spectrometer. *J. Am. Soc. Mass Spectrom.* **9**: 569–579.

31. Sobott, F., Hernandez, H., McCammon, M. G., Tito, M. A., and Robinson, C. V. (2002). A tandem mass spectrometer for improved transmission and analysis of large macromolecular assemblies. *Anal. Chem.* **74**: 1402–1407.

32. Chernushevich, I. V., and Thomson, B. A. (2004). Collisional cooling of large ions in electrospray mass spectrometry. *Anal. Chem.* **76**: 1754–1760.

33. Chernushevich, I. V., Loboda, A. V., and Thomson, B. A. (2001). An introduction to quadrupole-time-of-flight mass spectrometry. *J. Mass Spectrom.* **36**: 849–865.

34. Wang, Y., Schubert, M., Ingendoh, A., and Franzen, J. (2000). Analysis of non-covalent protein complexes up to 290 kDa using electrospray ionization and ion trap mass spectrometry. *Rapid Commun. Mass Spectrom.* **14**: 12–17.

35. Loo, J. A., Berhane, B., Kaddis, C. S., Wooding, K. M., Xie, Y., Kaufman, S. L., and Chernushevich, I. V. (2005). Electrospray ionization mass spectrometry and ion mobility analysis of the 20S proteasome complex. *J. Am. Soc. Mass Spectrom.* **16**: 998–1008.

36. Cheng, X., Chen, R., Bruce, J. E., Schwartz, B. L., Anderson, G. A., Hofstadler, S. A., Gale, D. C., Smith, R. D., Gao, J., and Sigal, G. B. (1995). Using electrospray ionization FTICR mass spectrometry to study competitive binding of inhibitors to carbonic anhydrase. *J. Am. Chem. Soc.* **117**: 8859–8860.

37. Loo, J. A. (2000). Electrospray ionization mass spectrometry: a technology for studying noncovalent macromolecular complexes. *Int. J. Mass Spectrom.* **200**: 175–186.

38. Olsen, J. V., Godoy, L. M. F. D., Li, G., Macek, B., Mortensen, P., Pesch, R., Makarov, A., Lange, O., Horning, S., and Mann, M. (2005). Parts per million mass accuracy on an Orbitrap mass spectrometer via lock mass injection into a C-trap. *Mol. Cell. Proteomics* **4**: 2010–2021.

39. Myung, S., Wiseman, J. M., Valentine, S. J., Takats, Z., Cooks, R. G., and Clemmer, D. E. (2006). Coupling desorption electrospray ionization with ion mobility/mass spectrometry for analysis of protein structure: evidence for desorption of folded and denatured states. *J. Phys. Chem. B* **110**: 5045–5051.

40. Bernstein, S. L., Liu, D., Wyttenbach, T., Bowers, M. T., Lee, J. C., Gray, H. B., and Winkler, J. R. (2004). Alpha-synuclein: stable compact and extended monomeric structures and pH dependence of dimer formation. *J. Am. Soc. Mass Spectrom.* **15**: 1435–1443.

41. Colgrave, M. L., Bramwell, C. J., and Creaser, C. S. (2003). Nanoelectrospray ion mobility spectrometry and ion trap mass spectrometry. *Int. J. Mass Spectrom.* **229**: 209–216.

42. Bacher, G., Szymanski, W. W., Kaufman, S. L., Zollner, P., Blaas, D., and Allmaier, G. (2001). Charge-reduced nano electrospray ionization combined with differential mobility analysis of peptides, proteins, glycoproteins, noncovalent protein complexes and viruses. *J. Mass Spectrom.* **36**: 1038–1052.

43. Sharon, M., Witt, S., Felderer, K., Rockel, B., Baumeister, W., and Robinson, C. V. (2006). 20S Proteasomes have the potential to keep substrates in store for continual degradation. *J. Biol. Chem.* **281**: 9569–9575.

44. Loo, J. A., Hu, P., McConnell, P., and Mueller, W. T. (1997). A study of Src SH2 domain protein–phosphopeptide binding interactions by electrospray ionization mass spectrometry. *J. Am. Soc. Mass Spectrom.* **8**: 234–243.

45. Lim, H.-K., Hsieh, Y. L., Ganem, B., and Henion, J. (1995). Recognition of cell-wall peptide ligands by vancomycin group antibiotics: studies using ion spray mass spectrometry. *J. Mass Spectrom.* **30**: 708–714.

46. Ayed, A., Krutchinsky, A. N., Ens, W., Standing, K. G., and Duckworth, H. W. (1998). Quantitative evaluation of protein–protein and ligand–protein equilibria of a large allosteric enzyme by electrospray ionization time-of-flight mass spectrometry. *Rapid Commun. Mass Spectrom.* **12**: 339–344.

47. Greig, M. J., Gaus, H., Cummins, L. L., Sasmor, H., and Griffey, R. H. (1995). Measurement of macromolecular binding using electrospray mass spectrometry. Determination of dissociation constants for oligonucleotide–serum albumin complexes. *J. Am. Chem. Soc.* **117**: 10765–10766.

48. Sannes-Lowery, K. A., Griffey, R. H., and Hofstadler, S. A. (2000). Measuring dissociation constants of RNA and aminoglycoside antibiotics by electrospray ionization mass spectrometry. *Anal. Biochem.* **280**: 264–271.

49. Zhang, S., Pelt, C. K. V., and Wilson, D. B. (2003). Quantitative determination of noncovalent binding interactions using automated nanoelectrospray mass spectrometry. *Anal. Chem.* **75**: 3010–3018.

50. Gao, J., Cheng, X., Chen, R., Sigal, G. B., Bruce, J. E., Schwartz, B. L., Hofstadler, S. A., Anderson, G. A., Smith, R. D., and Whitesides, G. M. (1996). Screening derivatized peptide libraries for tight binding inhibitors to carbonic anhydrase II by electrospray ionization–mass spectrometry. *J. Med. Chem.* **39**: 1949–1955.

51. Wigger, M., Eyler, J. R., Benner, S. A., Li, W., and Marshall, A. G. (2002). Fourier transform–ion cyclotron resonance mass spectrometric resolution, identification, and screening of non-covalent complexes of Hck Src homology 2 domain receptor and ligands from a 324-member peptide combinatorial library. *J. Am. Soc. Mass Spectrom.* **13**: 1162–1169.

52. Jurchen, J. C., and Williams, E. R. (2003). Origin of asymmetric charge partitioning in the dissociation of gas-phase protein homodimers. *J. Am. Chem. Soc.* **125**: 2817–2826.

53. Collins, F. S., Green, E. D., Guttmacher, A. E., and Guyer, M. S. (2003). A vision for the future of genomics research. *Nature* **422**: 835–847.

54. Frazier, M. E., Johnson, G. M., Thomassen, D. G., Oliver, C. E., and Patrinos, A. (2003). Realizing the potential of the genome revolution: The Genomes to Life program. *Science* **300**: 290–294.

2

SOFTLY, SOFTLY—DETECTION OF PROTEIN COMPLEXES BY MATRIX-ASSISTED LASER DESORPTION IONIZATION MASS SPECTROMETRY

KEVIN M. DOWNARD

School of Molecular and Microbial Biosciences, The University of Sydney, Australia

2.1 Introduction

2.2 First Glimpses and the First-Shot Phenomenon

2.3 Matrix and Solution Criteria to Preserve Protein Complexes

2.4 Laser Fluence, Wavelength, and Ion Extraction

2.5 Preservation of Protein Complexes on Conventional MALDI Targets

2.6 Affinity Targets and Surfaces Coupled to MALDI

2.7 Conclusions

    References

2.1 INTRODUCTION

Matrix-assisted laser desorption ionization (MALDI) is considered a "soft" ionization technique by virtue of its ability to ionize small chemical through large biological molecules without their fragmentation. This property alone, however, does not automatically extend to studies of noncovalent complexes that are far more susceptible to dissociation into their constituent molecules. Intermolecular noncovalent bonds can be dissociated at energies above some 20 kJ/mole, or approximately 1/20th of that required to cleave a covalent bond. The preservation of such complexes throughout sample deposition and the MALDI event, as well as during the extraction of ions from the source, flight through the mass analyzer, and detection through impact with the ion detector, clearly then requires a softly, softly approach. Intermolecular noncovalent protein–protein interactions are usually stabilized through multiple interactions between two or more protein molecules. These interactions take the form of ionic bonds, dipole–dipole and hydrogen bonds, hydrophobic and van der Waals forces. Collectively, such protein interactions are vital in biology in order to regulate biochemical pathways, control signaling, regulate protein function and assembly, and protect proteins from molecular damage by, for example, shielding proteins from radical attack and oxidative stress.

An attractive aspect of matrix-assisted laser desorption ionization mass spectrometry (MALDI-MS), which has advanced its use in many applications, is the relative ease with which samples can be prepared and analyzed on today's modern mass spectrometers. The approach is more tolerant of salts, buffers, denaturants, and other contaminants (Table 2.1) than electrospray ionization (ESI) [2]. It avoids the constraints imposed in solubilizing the sample in an appropriate solvent system, as in the case of ESI, and can produce a stable ion signal until the plated sample is consumed. These features belie the fact that the desorption and ionization of analytes during MALDI is actually a complex process involving optical phenomena and physicochemical and thermodynamic events, all of which are critical to successful analysis.

The complete MALDI-MS process encompasses several steps including sample preparation and deposition on the target surface with matrix, laser

**TABLE 2.1   Concentration of Denaturants, Buffers, and Salts
Tolerated by MALDI-MS**

Agent	Approximate Maximum Concentration
Alkali metal salts	0.5 M
Ammonium bicarbonate	0.05 M
Detergent (other than SDS)	0.1% (w/v)
Dithiothreitol	0.5 M
Guanidine HCl	0.5 M
Phosphate buffer	0.01 M
Sodium dodecylsuplhate	0.01% (w/v)
Tris HCl	0.05 M

Adapted from Mock et al. [1].

ablation and desorption of the sample analyte from the target, its electronic excitation, the generation of charged species in the gas phase usually via protonation or deprotonation, and finally the extraction, separation, and detection of those ions according to their mass-to-charge (m/z) ratio within the mass analyzer and detector, respectively. The development of MALDI [3] followed earlier "neat" (or analyte only) laser desorption studies. It involved the recognition that a dramatic improvement in ion production and thus sensitivity as well as the ionization of large macromolecules such as proteins could be attained if the analyte was mixed and deposited with a large mole excess of a specific "matrix" compound. The original matrix for use with an ultraviolet laser operating at 266 nm was nicotinic acid, which absorbs light efficiently at this wavelength [4]. These experiments were conducted at the same time that an alternative matrix, namely, glycerol containing a fine nickel powder, was reported [5, 6], earning the lead author a share of the Nobel Prize in Chemistry in 2002. With the proper combination of laser wavelength and matrix, a protein can be ionized and its ions detected in a mass spectrometer [7]. This latter approach, however, is far less practiced today than the approach originally reported by Karas and Hillenkamp [3, 4] and is therefore not described further here.

A number of mechanisms and models have been reported to describe the MALDI process [8–10] in terms of ablation [11] and desorption events [12] and in plume chemistries [13, 14]. At least in part, the protonation of molecules, in positive ion MALDI, has been shown to occur in the gas phase above the sample surface [15]. Computer simulations of these events have helped shed light on laser ablation processes [16].

It has been recognized from the outset that, in order to successfully detect noncovalent protein complexes by MALDI-MS, particular sample preparation, deposition, and ablation conditions are necessary. The acidic and typically solid-state sample environment of the deposited sample, combined with other features of the desorption/ionization and ion extraction process, was intuitively thought to adversely impact on the preservation of noncovalently bound complexes, including macromolecular protein complexes. Indeed, noncovalent complexes were initially detected infrequently and their detection was often transient and irreproducible. Evidence began to mount, however, that the preservation and detection of noncovalent complexes of proteins and other macromolecular ions was possible and indeed informative providing a number of experimental conditions and parameters were observed. If further developed and understood, MALDI-MS would offer many advantages over current analytical and physical techniques for studying protein interactions including the speed of analysis, the ability to quickly measure the molecular mass of complexes and assemblies to establish their stoichiometry, and to do so without the need to pretreat, tag, immobilize, or otherwise handle or manipulate proteins under investigation. The ability of mass spectrometers to resolve components of mixtures by mass, allows mixtures to be studied without the purification of protein components or their complexes.

This chapter assesses the state-of-play of MALDI mass spectrometry for the study of protein complexes since an early review on the subject [17]. It presents the documented experimental conditions required to preserve and maintain protein interactions throughout sample preparation and analysis, and the information that can be gleaned from such experiments. This includes the stability of such complexes, the stoichiometry of protein multimers and complexes, the nature and site of interaction interfaces and their biological significance, and the use of preformed surfaces and affinity-based methods in conjunction with MALDI-MS that are known by a range of names and acronyms. Such studies add to the repertoire of mass spectrometric approaches for studying protein complexes [18], many of which are described elsewhere in this book.

2.2   FIRST GLIMPSES AND THE FIRST-SHOT PHENOMENON

Evidence of at least the transient existence of ions of protein multimers first appeared in the literature shortly after the development of the MALDI approach. Multimeric forms of the proteins streptavidin [19] and glucose isomerase [20] were reported in 1990, but the first investigations focused on the specific preservation and detection of protein and peptide complexes by MALDI-MS appeared five years later [21, 22].

A "first-shot" phenomenon was reported in the early protein complex studies [21, 23] where protein aggregates were only seen following the first laser pulse; all subsequent laser shots at the same target position yielded primarily protein monomer. In the case of the association of subunits of the hydrophilic protein streptavidin [21, 23], a tetrameric protein ion (Q^+) was observed as the base peak following the first laser ablation of the surface consistent with its native form. A dimer of this tetramer ($2Q^+$) was also observed, albeit of much lower intensity, together with ions of some monomer (M^+) and dimeric (D^+) proteins. Subsequent analysis of the same target position produced only protein monomer (Figure 2.1). A possible explanation for this phenomenon is either that protein complexes precipitate from the droplet applied to the target surface, and subsequently reside on the top of the matrix after drying, and/or that surface is depleted of matrix over the interior of the deposited sample where it destabilizes and dissociates the complex. Thus, in one model to explain this phenomenon, it is proposed that a disproportionate amount of intact complex resides on the sample surface and is ablated from that level only. Yet as has been seen in later studies, the

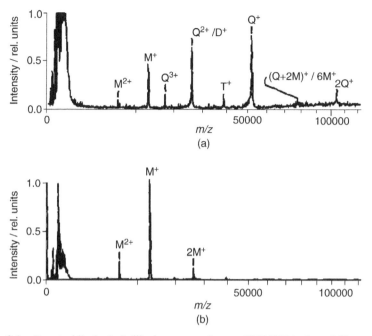

Figure 2.1   Streptavidin in 2, 6-dihydroxyacetophenone/CAN/TFA. λ = 3 55 n m. M = monomer, D = dimer, T = trimer, Q = tetramer. (a) First shot only (30 spectra total), (b) 25 shots [23].

first-shot phenomenon is not always evident, so that its importance should not be overstated.

A systematic investigation of the ability to probe subunit stoichiometry of protein complexes was reported by Moniatte and co-workers [24]. Noncovalent complexes of aerolysin, α-hemolysin, and the enzymes bovine liver catalase, pyruvate k inase, and a lcohol de hydrogenase, a nd the lectin concanavalin A were studied. The detection of specific noncovalent complexes was found to be dependent on the choice of matrix, the sample preparation procedure, and the speed of evaporation of solvent after deposition of the sample. A t ime-dependent study examined the formation of a heptameric complex of aerolysin over a 12 hour period by MALDI-MS following deposition of sample using a two-step, s andwich depos ition me thod, wi th s ample depos ited neat o nto a thin fi lm o f mat rix a nd co vered with t he s ame (Figure 2 .2). A hep tameric complex of α-hemolysin was also observed in accord with crystallographic data of this stoichiometry [25] that could not detected by ESI-MS. The fast evaporation o f s olvent w as re quired i n t he ca se o f p H-sensitive co mplexes where m ixed mat rix–sample s olutions w ere depos ited o nto t he p late. T his minimizes the time the complex is in contact with the matrix and the acidic solution environment.

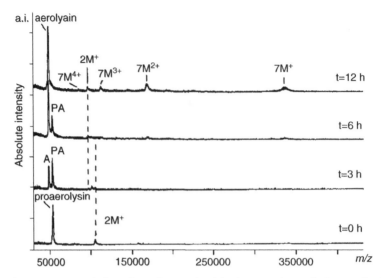

Figure 2.2   MALDI-MS time-dependent study of the heptameric aerolysin complex formation during the tryptic activation of proaerolysin. The heptamer is clearly observed after 12 h
incubation. Measured molecular mass of the heptamer: 334,900 Da. M designates monomer,
2M dimer, 7M aerolysin heptamer, PA proaerolysin, A aerolysin. Nonspecific dimers are
detected at 3 h, while specific intermediates are visible at 12 h [24].

2.3   MATRIX AND SOLUTION CRITERIA TO PRESERVE PROTEIN COMPLEXES

By definition, the matrix is an indispensable component of the MALDI process. It serves a number of different, but essential, functions that include the
direct absorption of the laser light. This in turn helps prevent the thermal and
photochemical destruction of the analyte. In the case of the ionization of analytes in the positive ion mode, molecules typically receive a proton from the
matrix molecules. Successful MALDI matrix compounds [4, 26–28] therefore tend to be at least mildly acidic, with carboxylic acid and/or phenolic
functional groups, in addition to their conjugated and/or aromatic character
that provides their chromophoric properties (Table 2.2). Furthermore, it is
common practice to prepare matrix solutions in which other acids such as
acetic acid and trifluoroacetic acid are added to promote the ionization process. These, however, should be avoided where protein complexes are to be
maintained and detected [29]. A compromise is struck between maintaining
an acidic environment to effect or promote ionization, and one closer to a
physiological pH required to preserve many protein complexes in solution.

The nature of the matrix and solvent used for the mixed matrix/sample solutions has been shown to greatly impact on the detection of noncovalent protein
complexes. Intact steptavidin tetramer was observed for matrices of ferulic

TABLE 2.2 Common UV MALDI Matrices and Preparation Solvent [26–28]

Compound	Symbol (Name)	[M+H]+ Monoisotopic m/z	Structure	Solvent	Wavelength (nm)
α-Cyano-4-hydroxy-cinnamic acid	aCCA	190.05		Acetonitrile, water, ethanol, acetone	337
3,5-Dimethoxy-4-hydroxy-cinnamic acid	SA (sinapinic acid)	225.07		Acetonitrile, water, acetone, chloroform	337, 266
2,5-Dihydroxy benzoic acid	DHB	155.03		Acetonitrile, water, acetone, methanol, chloroform	337, 266
2,6-Dihydroxy acetophenone	DHAP	153.06		Acetonitrile, ethanol, tetrahydrofuran	337, 266
4-Hydroxy-3-methoxy-cinnamic acid	FA (ferulic acid)	195.07		Acetonitrile, water, propanol	337, 266

acid and isomers of dihydroxyacetophenone (DHAP), although the monomeric form of the protein dominated when trihydroxyacetophenone (THAP) and 2,5-dihydroxybenzoic acid (DHB) were used [23]. Similar results were obtained for yeast alcohol dehydrogenase and beef liver catalase with quarternary complexes detected for both using matrix DHAP [21]. The dilution of analyte molecules in solution and on the target can prevent or disrupt protein aggregation and complex preservation and the transfer of energy in the gas phase above the target together with proton transfer to effect the ionization event.

Matrix acidity appears to be a factor in these results with 2,5-DHB more acidic (pK_a 2.9) than ferulic acid (pK_a 4.5). Jesperson et al. [30] have reported on matrix acidity effects for the successful detection of complexes of steptavidin, glutathione transferase, and hemoglobin after their deposition in aqueous solutions with matrix 3-hydroxypicolinic acid at pH 3.8 over both more acidic and basic solution environments. No first-shot phenomenon was evident in this study.

The order of matrix and sample deposition can also aid in the preservation and ultimate detection of protein complexes with a two-step approach in which the addition of mixed sample and matrix to a preformed dry matrix film gave superior results in one report [24]. It is also common practice to prepare matrix solutions in organic solvents such as acetonitrile, acetone, chloroform, common alcohols, and tetrahydrofuran (see Table 2.2). The presence of organic solvents helps to assist with the evaporation of the plated analyte–matrix solution and control crystal growth, homogeneity, and spread of the sample on the target surface. As stated previously, while these solvents can minimize the time protein complexes spend in the matrix solution, they are denaturing and may also play a role in dissociating protein complexes prior to their analysis.

There is as yet no consensus or common sample preparation that has been found suitable for studying protein complexes by MALDI mass spectrometry. The sample preparation procedures described previously have largely been derived by trial and error. It is unlikely too that a common approach for studying protein complexes by MALDI will be found, given the dramatic differences apparent in the stability and nature of the wide variety of protein interactions known to date. There are, nonetheless, important criteria to observe to attempt to preserve protein complexes during sample preparation and deposition, as described previously, and during the irradiation and extraction of ions from MALDI targets as reviewed in the next section.

2.4   LASER FLUENCE, WAVELENGTH, AND ION EXTRACTION

While the matrix and sample solution conditions are important in order to sustain a noncovalent complex during its isolation or formation, and during its deposition onto the target surface, the manner in which the complex is ionized in, and extracted from, the ion source is critical to its detection.

Rosinke et al. [21] reported that the generation of noncovalent complexes strongly depends on the laser fluence. In the case of the protein porin, its trimer is seen at 111.25 kDa. at low laser power. With increasing irradiance, above the threshold at which ions are seen, ions of the monomer subunit at 37.1 kDa begin to dominate the spectrum at the expense of the trimer (Figure 2.3). Where a complex is disproportionately located at the surface, which has been suggested to explain the first-shot phenomenon, it is more susceptible to direct laser irradiation. When laser radiation strikes a surface, the laser energy is transformed to heat. The temperature of a solid will then increase, leading to the potential melting and evaporation or sublimation of the material. The local heating in the vicinity of an embedded complex sample has the potential to dissociate the complex and/or effect molecular damage. Yet the ability of small and large molecules to resist degradation during MALDI-MS, and be detected as molecular ions over fragments, suggests little heat is transferred to the sample analyte itself. In this respect, the matrix appears to insulate the sample molecules from local heating.

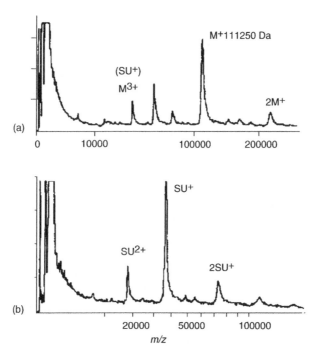

Figure 2.3 UV-MALDI mass spectra of porin using matrix ferulic acid in THF. (a) Sum of first shots onto a given spot; (b) after multiple shots [21].

In a typical UV-MALDI-MS experiment, the laser fluence is some 10–100 mJ/cm^2. Only a fraction of this energy is transferred to the target over a typical irradiation diameter of 10–100 μm. It is difficult to measure the temperature increase at the irradiated spot on the target, and this is influenced by the nature and the size of the crystals [31], but it has been estimated that the maximum surface temperature reached for a 355 nm laser pulse of 20 mJ/cm^2 is 550 K [8]. For the threshold desorption of neutral molecules from a 200 μm spot, a temperature of 464 K or 191°C has been reported [32]. This heating under vacuum can lead to sublimation of the matrix and sample, even at laser powers below the threshold at which ions are detected. The sudden input of energy from the laser into the sample can also lead to sample vaporization rather than pure sublimation [33].

An infrared (IR) over ultraviolet laser may afford some advantages for studying noncovalent macromolecular complexes. To overcome the high sample consumption per laser shot found to be required using organic matrices, liquid matrices such as glycerol and even ice have been shown to be a suitable matrix for IR-MALDI. Results are comparable to experiments with solid matrices using an ultraviolet laser [34]. Liquid matrices offer a more homogeneous sample morphology with more sample replenished at the irradiation site after some is consumed, in common with fast atom bombardment (FAB) ionization [35], thus offering a high shot-to-shot signal stability. Avoiding the acidic and organic solution environments common to UV-MALDI experiments offers obvious benefits in terms of the preservation and analysis of protein complexes.

At low wavelengths in the IR of 3 μm, a lower threshold fluence is required for successful ionization and detection of analytes over UV-MALDI resulting in minimal fragmentation of analytes [36], even those exceeding 100 kDa. The velocity and energy distribution of ions released is, however, broader and this leads to poorer mass resolution in IR-MALDI spectra over experiments performed in the UV. Ion production is also usually poorer in IR-MALDI experiments, particularly for solid matrices, leading to lower sensitivities. This, and the higher cost of IR lasers over nitrogen (337 nm) and Nd-YAG (typically 266 μm) lasers, has impeded the application of IR-MALDI [12].

Although IR-MALDI is reported to be a softer ionization approach [37], in terms of energy transfer [38], there is as yet no comprehensive study to illustrate the benefits of the MALDI conducted at wavelengths in the infrared for the study of protein complexes. Double-stranded deoxyribonucleic acid (DNA), however, was successfully desorbed and detected by IR-MALDI-MS using a mixed matrix comprising glycerol and ammonium acetate or tris(hydroxymethyl)aminomethane hydrochloride. The presence of salt was found to enhance the formation and detection of the DNA dimer, and the laser power was also critical to detection, where an increase in fluence by just 1.2-fold above the threshold for detection resulted in the dissociation of the complex.

An additional issue that impacts on the detection of molecular ions in general, and less stable noncovalent complexes in particular, is their extraction from the ion source. Ions typically extracted from a MALDI ion source under acceleration potentials of 20–30 kV and their singly charged ions therefore have energies of 20–30 keV (or 300–500 µJ). Subject to the pressures of the ion source and mass analyzer, such high-energy ions are prone to dissociate through collision with residual gas molecules as they pass through a mass spectrometer at high velocities. Molecular ions are ejected from the MALDI plume at speeds of some 500–1000 ms^{-1}. Noncovalent bonds are particularly susceptible in this regard, such that an ionized noncovalent complex that survives deposition and ablation from the MALDI surface may still dissociate into its constituents on the way to the detector. It is likely then that only a smaller population of the ions formed or maintained in the ion source are ultimately detected, except in the case of extremely stable complexes.

2.5 PRESERVATION OF PROTEIN COMPLEXES ON CONVENTIONAL MALDI TARGETS

It has been shown that peptide antigen–antibody interactions can survive on a conventional MALDI surface while other nonbinding peptides are ionized from it, providing another means with which to probe protein interactions. Results for protein lysozyme [29], and the hemagglutinin antigen of a strain of the influenza virus, in a subsequent mixed viral protein study [39], have shown that the solution-specific associations between a monoclonal antibody and their target antigens can be preserved on the MALDI target after deposition in a α-cyano-4-hydroxycinnamic acid matrix solution at a pH ~3 (Figure 2.4). When trifluoroacetic acid is added to the matrix solution, lowering its pH, dissociation of the complex occurs, resulting in a subsequent increase in the relative area and detection of ions for the binding peptide in the MALDI spectrum of the antibody-treated mixture [29]. This indirect method of studying protein complexes by MALDI requires a comparison of a control (no antibody) MALDI spectrum with that of the antibody-treated mixture. The binding peptide(s) is/are determined from the decrease in absolute area of its ions across both spectra relative to a nonbinding peptide signal or the base peak. An algorithm has been developed to aid with the identification of the binding peptide in these digests or other more complex mixtures [40]. In the case of a mixed antigen study in which a whole virus digest was treated with a monoclonal antibody to one surface antigen, an epitope of the hemagglutinin antigen of type A influenza strains was characterized [39]. Direct

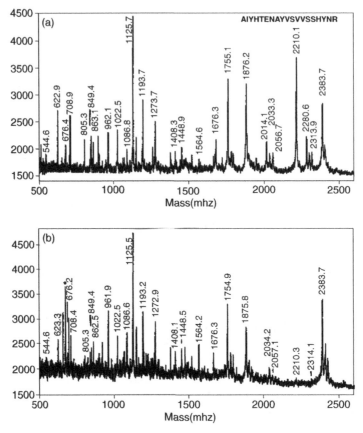

Figure 2.4 MALDI mass spectra of the tryptic digest of the viral proteins from a type A influenza strain (a) before and (b) after reaction with monoclonal antibody [39].

confirmation of the validity of this result came from the subsequent detection of the peptide epitope–antibody complex by MALDI-MS. Using different extraction conditions, it was also possible to desorb the entire peptide antigen–antibody complex from the MALDI target and detect it as a series of low multiply charged ions [41]. Such experiments, in which nonbinding peptide or protein partners serve as a control, allow one to more confidently assess the preservation or detection of a specific noncovalent protein complex.

The same approach employed to identify antigenic determinants or epitopes of protein antigens [29, 39] was applied to study protein–ligand complexes though redefined as "intensity-fading" experiments [42]. In these experiments, relative intensities of protease inhibitors were compared with a nonbinding control in the presence and absence of protease, illustrating that these noncovalent complexes can also survive on the MALDI target in the matrix sinapinic acid deposited in a solution of acetonitrile and water.

In order to extend the approach to biological mixtures as part of a proteomics strategy for analyzing protein interactions, the earlier immunoaffinity MALDI studies [29, 39] have recently been extended to gel-recovered antigens [43]. In these studies, in-gel digested protein was recovered and one portion treated with a monoclonal antibody raised to the antigen. Epitopic peptides were identified in accord with earlier work [39] with greater coverage evident in the mass map for gel-purified antigens [43] over mixed antigen, whole virus work [39]. This improves the probability that the binding peptide will be detected in the control (no antibody) spectrum, and its relative ion signal decrease followed in the antibody-treated mixture. In the study of whole virus digests and other biological extracts, different ionization efficiencies and suppression effects can result in the absence of some components in a MALDI spectrum, preventing its study in this manner. Gel separation of components prior to their treatment and analysis overcomes this.

The proteomics approach reported in the antigenic surveillance of the influenza virus has utility for the study of any protein complex either studied after its direct recovery from a native gel in an intact form [44], or formed after recovery of a constituent protein and treated with an interacting protein, ahead of MALDI analysis.

2.6 AFFINITY TARGETS AND SURFACES COUPLED TO MALDI

In an approach first described as surface-enhanced affinity capture (SEAC) [45], and what is now more widely known as surface-assisted (SALDI) or surface-enhanced laser desorption ionization (SELDI), protein–protein interactions from biological extracts can be studied [46, 47]. Protein binding partners are bound to common chromatographic supports or specifically designed affinity media. Biological extracts are then added to the surface and the unbound proteins and other interfering compounds washed off. The proteins bound through protein–protein interactions are retained and then desorbed directly from the surface by laser ablation with subsequent analysis and detection MALDI-MS. Reviews of the SELDI approach [48, 49], which has been commercialized [50] and particularly promoted for biomarker discovery in cancer research [51], have been reported. Similar surface-based approaches have been exploited to study phosphoproteins in human saliva. An alkaline phosphatase-bioreactive probe, in which the enzyme was covalently bound to an activated gold-coated mass spectrometry target, has been applied for the recovery and subsequent identification of phosphoproteins.

Lehman et al. [47] detected specific interactions among S100 proteins using anti-S100A8 antibody coupled on immunoaffinity beads. Such affinity-based

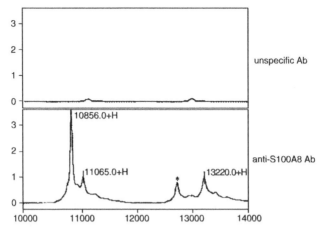

Figure 2.5 Protein–protein interaction assay using specific anti-S100A8 antibody coupled with SELDI-MS analysis. The ion signals at 10.87 kDa, 11.06 kDa, and 13.22 kDa represent S100A8, S 100A10, a nd S 100A9, r espectively. T he s ignal la beled b y t he a sterisk w as not identified [47].

methods have their roots in mass spectrometric immunoassay reported for the parallel a nalysis of a ffinity-captured t arget a ntigens b y M ALDI ma ss spectrometry [52] in an approach that has been applied for the direct extraction and quantification of insulin growth factors (IGFs) from human plasma [53] to the low pM level. In the work of Lehman and colleagues, the antibody-coupled beads were incubated with immortalized human keratinocyte (HaCaT) cells. Proteins eluted from both anti-S100A8 immobilized antibody, and a nonspecific antibody control, were digested with trypsin and the peptides were analyzed by SELDI-MS on a reverse-phase NP20 array. An interaction partner identified as calpactin light chain (S100A10) with a m olecular mass of 11.07 kDa was identified (Figure 2.5). Its i nteraction with S 100 b inding w as co nfirmed by coimmunoprecipitation experiments.

The coupling of the surface plasmon resonance (SPR) approach with the analysis o f pro tein i nteraction pa rtners a fter t heir l aser-induced re lease by MALDI-MS has been described as biomolecular interaction analysis (BIA) mass spectrometry [54–56]. Proteins recovered on the SPR sensor chip can be analyzed directly by MALDI-MS ("on-chip"), after the addition of a matrix, o r following e lution a nd m icrorecovery [57]. T he pro tein co mplex between a retinol binding protein (RBP) and transthyretin (TTR) was detected after recovery from human plasma passed over the surface of sensor chips derivatized with antibodies to either protein. A MALDI-TOF mass spectrum taken for proteins bound to t he chips revealed signals from both R BP and TTR, indicating retrieval of the RBP–TTR protein complex (Figure 2.6).

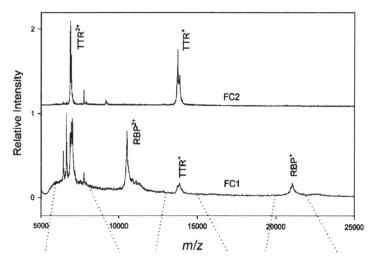

Figure 2.6   MS analysis from the surfaces FC1 derivatized with anti-RBP and FC2 derivatized with anti-TTR following the SPR analysis of human plasma providing evidence for the RBP–TTR complex [58].

2.7   CONCLUSIONS

Although less developed than ESI-MS (described in Chapter 1 of this book) for t he d irect s tudy o f pro tein i nteractions, t he ra nge o f a pplications a nd approaches employing MALDI-MS demonstrates that it too has a role to play in their analysis. The simplicity, speed of analysis, and low detection limits afforded by M ALDI-MS, coupled with the ab ility to s tudy pro tein i nteractions from cellular extracts, plasma, a nd sera, make it an attractive a lternative to other analytical and spectroscopic approaches. Like ESI-based studies, a de gree o f t rial a nd er ror, h owever, i s s till re quired i n t he a pplication o f MALDI-MS for this task with no common protocol currently available that is applicable for all protein systems. A range of conditions and protocols have been reported t hat ha ve yielded s uccessful res ults for s pecific applications. These offer promise and encouragement for future work that should cement the role of MALDI-MS for investigating protein interactions.

REFERENCES

1. Mock, K. K., Sutton, C. W., and Cottrell, J. S. (1992). Sample immobilization protocols for matrix-assisted laser-desorption mass spectrometry. *Rapid Commun. Mass Spectrom.* **6**: 233–238.

2. Coligan, J. E., D unn, B. M., a nd P loegh, H. L . (1995). Mat rix-assisted la ser desorption/ionization t ime-of-flight mass analysis of peptides. In *Current*

Protocols in Protein Science (D.W. Speicher and P.T. Wingfield, Eds.), Vol. 1, John Wiley & Sons, Hoboken, NJ.

3. Karas, M., Bachmann, D., Bahr, U., and Hillenkamp, F. (1987). Matrix-assisted ultraviolet laser desorption of non-volatile compounds. *Int. J. Mass Spectrom. Ion Proc.* **78**: 53.

4. Karas, M., and Hillenkamp, F. (1988). Laser desorption ionization of proteins with molecular masses exceeding 10000 daltons. *Anal. Chem.* **60**: 2299–2301.

5. Tanaka, K., Ido, Y., Akita, S., Yoshida, Y., and Yoshida, T. (1987). Detection of high mass molecules by laser desorption time-of-flight mass spectrometry. *Proceedings of the Second Japan–China Joint Symposium on Mass Spectrometry*, pp. 185–188.

6. Tanaka, K., Waki, H., Ido, Y., Akita, S., Yoshida, Y., and Yoshida, T. (1988). Protein and polymer analyses up to m/z 100 000 by laser ionization time-of-flight mass spectrometry. *Rapid Commun. Mass Spectrom.* **2**: 151–153.

7. Tanaka, K. (2003). The origin of macromolecule ionization by laser irradiation, *Nobel Prix* **2002**: 197–217.

8. Knochenmuss, R. (2002). A quantitative model of ultraviolet matrix-assisted laser desorption/ionization. *J. Mass Spectrom.* **37**: 867–877.

9. Georgiou, S., and Hillenkamp, F. (2003). Laser ablation and molecular substrates. *Chem. Rev.* **103**: 317–319.

10. Karas, M., and Krüger, R. (2003). Ion formation in MALDI: the cluster ionization mechanism. *Chem. Rev.* **103**: 427–440.

11. Paltauf, G., Dyer, P. E. (2003). Photomechanical processes and effects in ablation. *Chem. Rev.* **103**: 487–518.

12. Dreisewerd, K. (2003). The desorption process in MALDI. *Chem. Rev.* **13**: 395–426.

13. Frankevich, V., Zhang, J., Friess, S. D., Dashtiev, M., and Zenobi, R. (2003). Role of electrons in laser desorption/ionization mass spectrometry. *Anal. Chem.* **22**: 6063.

14. Breuker, K., Knochenmuss, R., Zhang, J., Stortelder, A., and Zenobi, R. (2003). Thermodynamic control of final ion distributions in MALDI: in-plume proton transfer, reactions. *Int. J. Mass Spectrom.* **216**: 21.

15. Wang, B. H., Dreisewerd, K., Bahr, U., Karas, M., and Hillenkamp, F. (1993). Gas-phase cationization and protonation of neutrals generated by matrix-assisted laser desorption. *J. Am. Soc. Mass Spectrom.* **4**: 393–398.

16. Zhigilei, L. V., Leveugle, E., Garrison, B. J., Yingling, Y. G., Zeifman, M. I., (2003). Computer simulations of laser ablation of molecular substrates. *Chem. Rev.* **103**: 321–348.

17. Farmer, T. B., and Caprioli, R. M. (1998). Determination of protein–protein interactions by matrix-assisted laser desorption/ionization mass spectrometry. *J. Mass Spectrom.* **33**: 697–704.

18. Downard, K. M. (2006). Ions of the interactome — the role of mass spectrometry in the study of protein interactions in proteomics and structural biology. *Proteomics*, **6**: 5374–5384.

19. Karas, M., Bahr, U., Ingendoh, A., Nordhoff, E., Stahl, B., Strupat, K., and Hillenkamp, F. (1990). Principles and applications of matrix-assisted UV laser desorption ionization mass spectrometry. *Anal. Chim. Acta* **241**: 175–185.

20. Karas, M., and Bahr, U. (1990). Laser desorption ionization mass spectrometry of large biomolecules. *Trends Anal. Chem.* **9**: 321–325.

21. Rosinke, B., Strupat, K., Hillenkamp, F., Rosenbusch, J., Dencher, N., Kruger, U., and Galla, H.-J. (1995). Matrix-assisted laser desorption/ionization mass spectrometry (MALDI-MS) of membrane proteins and non-covalent complexes, *J. Mass Spectrom.* **30**: 1462–1468.

22. Woods, A. S., Buchsbaum, J. C., Worrall, T. A., Berg, J. M., Cotter, R. J. (1995). Matrix-assisted laser desorption ionization of non-covalently bound compounds. *Anal. Chem.* **67**: 4462–4465.

23. Cohen, L. R. H., Strupat, K., and Hillenkamp, F. (1997). Analysis of quaternary protein ensembles by matrix assisted laser desorption/ionization mass spectrometry. *J. Am. Soc. Mass Spectrom.* **8**: 1046–1052.

24. Moniatte, M., Lesieur, C., Vecsy-Semjen, B., Buckley, J. T., Pattus, F., van der Goot, F. G., and Van Dorsselaer, A. (1997). Matrix-assisted laser desorption–ionization time-of-flight mass spectrometry in the subunit stoichiometry of high mass non-covalent complexes. *Int. J. Mass Spectrom.* **169/170**: 179–199.

25. Gouaux, J. E., Braha, O., Hobaugh, M. R., Song, L., Cheley, S., Shustak, C., and Bayley, H. (1994). Subunit stoichiometry of staphylococcal alpha-hemolysin in crystals and on membranes: a heptameric transmembrane pore, *Proc. Natl. Acad. Sci. USA* **91**: 12828–12831.

26. Beavis, R. C., Chait, B. T., and Standing, K. G. (1989). Matrix-assisted laser-desorption mass spectrometry using 355 nm radiation. *Rapid Commun. Mass Spectrom.* **3**: 436–439.

27. Beavis, R. C., Chaudhary, T., and Chait, B. T. (1992). α-Cyano-4-hydroxycinnamic acid as a matrix for matrix-assisted laser desorption mass spectrometry. *Org. Mass Spectrom.* **27**: 156–158.

28. Strupat, K., Karas, M., and Hillenkamp, F. (1991). 2,5-Dihydroxybenzoic acid: a new matrix for laser desorption–ionization mass spectrometry. *Int. J. Mass Spectrom. Ion Proc.* **111**: 89–102.

29. Kiselar, J. G., and Downard, K. M. (1999). Direct identification of protein epitopes by mass spectrometry without immobilization of antibody and isolation of antibody–peptide complexes. *Anal. Chem.* **17**: 1792–1801.

30. Jespersen, S., Niessen, W. M. A., Tjaden, U. R., and van der Greef, J. (1998). Basic matrices in the analysis of non-covalent complexes by matrix-assisted laser desorption/ionization mass spectrometry. *J. Mass Spectrom.* **33**: 1088–1093.

31. Sadeghi, M., and Vertes, A. (1998). Crystallite size dependence of volitilization in m atrix-assisted l aser d esorption i onization, *Appl. Surface Sci.* **127–129**: 226–234.

32. Dreisewerd, K., Schuerenberg, M., Karas, M., and Hillenkamp, F. (1995). Influence of the laser intensity and spot size on the desorption of molecules and ions in matrix-assisted laser desorption/ionization with a uniform beam profile. *Int. J. Mass Spectrom. Ion Proc.* **141**: 127–148.

33. Price, D. M., Bashir, S., and Derrick, P. R. (1999). Sublimation properties of *x,y*-dihydroxybenzoic a cid iso mers a s mo del m atrix-assisted la ser desor ption ionization (MALDI) matrices, *Thermochimica Acta.* **327**: 167–171.

34. Berkenkamp, S., Karas, M., and Hillenkamp, F. (1996). Ice as a matrix for IR-matrix-assisted laser desorption/ionization: mass spectra from a protein single crystal. *Proc. Natl. Acad. Sci. USA* **93**: 7003–7007.

35. Barber, M, Bordoli, R. S., Sedgwick, R. D., and Tyler, A. N. (1981). Fast atom bombardment of solids as an ion source in mass spectroscopy. *Nature* **293**: 270–275.

36. Berkenkamp, S. Menzel, C. Karas, M., and Hillenkamp, F. (1997). Performance of infrared matrix-assisted laser desorption/ionization mass spectrometry with lasers emitting in the 3 mm wavelength range. *Rapid Commun. Mass Spectrom.* **11**: 1399–1406.

37. Cramer, R., and Burlingame, A. L . (2000). I R-MALDI—softer i onization i n MALDI-MS for s tudies o f lab ile m acromolecules. I n *Mass Spectrometry in Biology and Medicine* (A. L. Burlingame, S. A. Carr, and M. A. Baldwin, Eds.), pp. 289–307, Humana Press, New Jersey.

38. Talrose, V. L., Person, M. D., Whittal, R. M., Walls, F. C., Burlingame, A. L., and Baldwin, M. A. (1999). Insight into absorption of radiation/energy transfer in infrared matrix-assisted laser desorption/ionization: the roles of matrices, water and metal substrates, *Rapid Commun. Mass Spectrom.* **13**: 2191–2198.

39. Kiselar, J. G., and Downard, K. M. (1999). Antigenic surveillance of the influenza virus by mass spectrometry. *Biochemistry* **43**: 14185–14191.

40. Ho, J. W. K., Morrissey, B., and Downard, K. M. (2006). A computer program for the identification of protein interactions from the spectra of masses (PRISM), *J. Am. Soc. Mass Spectrom.* **18**: 563–566.

41. Kiselar, J. G., and Downard, K. M. (2000). Preservation and detection of specific antibody–peptide complexes by matrix-assisted laser desorption ionization mass spectrometry. *J. Am. Soc. Mass Spectrom.* **11**: 746–750.

42. Villanueva, J., Yanes, O., Querol, E., Serrano, L., and Aviles, F. X. (2 003). Identification of protein ligands in complex biological samples using intensity-fading MALDI-TOF mass spectrometry. *Anal. Chem.* **75**: 3385–3395.

43. Morrissey, B., and Downard, K. M. (2 006). A prot eomics approa ch to s ur-vey the antigenicity of the influenza virus by mass spectrometry. *Proteomics.* **6**: 2034–2041.

44. Mackun, K., and Downard, K. M. (2003). Strategy for identifying protein–protein interactions o f g el-separated prot eins a nd co mplexes b y m ass sp ectrometry. *Anal. Biochem.* **318**: 60–70.

45. Hutchens, T. W., and Yip, T. T. (1993). New desorption strategies for the mass spectrometric a nalysis of m acromolecule. *Rapid Commun. Mass Spectrom.* **7**: 576–580.

46. Schweigert, F. J ., W irth, K ., a nd R aila, J . (2 004). C haracterization o f t he microheterogeneity of transthyretin in plasma and urine using SELDI-TOF-MS immunoassay. *Proteome Sci.* **2**: 5.

47. Lehmann, R., Melle, C., Escher, N., and von Eggeling, F. (2005). Detection and identification of protein interactions of S100 proteins by ProteinChip technology. *J. Proteome Res.* **4**: 1717–1721.

48. Merchant, M. , a nd W einberger, S . R. (2 000). R ecent a dvancements i n s urface-enhanced la ser desor ption/ionization–time o f fl ight–mass spectrometry. *Electrophoresis* **21**: 1164–1177.

49. Tang, N., Tornatore, P., and Weinberger, S. R. (2004). Current developments in SELDI affinity technology. *Mass Spectrom. Rev.* **23**: 34–44.

50. Wright, G. L ., Jr. (2 002). S ELDI P roteinChip MS : a p latform f or b iomarker discovery and cancer diagnosis. *Expert Rev. Mol. Diagn.* **2**: 549–563.

51. Yip, T. T., and Lomas, L. (2002). SELDI ProteinChip array in oncoproteomic research. *Technol. Cancer Res. Treat.* **1**: 273–280.

52. Nelson, R. W ., K rone, J. R. , B ieber, A. L ., a nd W illiams, P. (1995). Ma ss-Spectrometric Immunoassay, *Anal. Chem.* **67**: 1153–1158.

53. Nelson, R. W., Nedelkov, D., Tubbs, K., and Kiernan, U. A. (2004). Quantitative mass spectrometric immunoassay of insulin like growth factor 1, *J. Proteome Res.* **4**: 851–855.

54. Krone, J. R. , N elson, R. W ., D ogruel, D., W illiams, P., G ranzow, R. (1997). BIA/MS: interfacing biomolecular interaction analysis with mass spectrometry. *Anal. Biochem.* **244**: 124–132.

55. Nelson, R. W ., N edelkov, D ., a nd T ubbs, K . A. (2 000). B iosensor c hip m ass spectrometry: a chip-based proteomics approach. *Electrophoresis* **21**: 1155–1163.

56. Nedelkov, D ., a nd N elson, R. W . (2 003). S urface p lasmon resona nce m ass spectrometry: recent progress and outlooks. *Trends Biotecnol.* **21**: 301–305.

57. Nedelkov, D., and Nelson, R. W. (2003). Delineating protein-protein interactions via biomolecular interaction analysis-mass spectrometry. *J. Mol. Recogn.* **16**: 9–14.

58. Nedelkov, D ., a nd N elson, R. W . (2 001). D elineation o f i n v ivo a ssembled multiprotein complexes via biomolecular interaction analysis mass spectrometry. *Proteomics* **1**: 1441–1446.

3

PROBING PROTEIN INTERACTIONS USING HYDROGEN–DEUTERIUM EXCHANGE MASS SPECTROMETRY

DAVID D. WEIS, SUMA KAVETI, YAN WU, AND JOHN R. ENGEN

Department of Chemistry, University of New Mexico, Albuquerque, New Mexico 87131

3.1 Introduction

3.2 Hydrogen Exchange Background

3.3 General HX-MS Method

    3.3.1 Location Information Provided by HX-MS

    3.3.2 Revealing Interactions by Comparison

3.4 Interactions of Proteins

3.5 Examples

    3.5.1 Conformational Changes of Proteins During Binding

    3.5.2 Protein–Protein Interactions

    3.5.3 Protein–Peptide Interactions

    3.5.4 Protein–Small Molecule Interactions

3.6 Conclusions

    Acknowledgments

    References

3.1 INTRODUCTION

Protein interactions are important. Robust methods that reveal the details of the interactions, in terms of their locations and the effects of the interactions on the conformation and dynamics of the proteins, are essential. One method that provides this level of detail is hydrogen–deuterium exchange mass spectrometry. Deuterium oxide (D_2O) is used as a probe of protein conformation and dynamics, and the incorporation of the deuterium is measured with mass spectrometry. A number of advantages present themselves: mixtures can be analyzed (including the protein of interest and its binding partner(s)), only small quantities of material are needed, and information can be localized to 5–20 amino acids stretches. Any size protein can be probed, dissociation constants can be obtained, and the regions altered by protein interactions can be located.

3.2 HYDROGEN EXCHANGE BACKGROUND

Hydrogen exchange (HX) is a phenomenon whereby labile hydrogens in proteins exchange with hydrogens in the surrounding solvent. The details of hydrogen exchange mechanisms have been widely reviewed [1–5] as has coupling hydrogen exchange with mass spectrometry (MS) [6–11]. A short summary based on these references is presented here.

Of the many different kinds of hydrogens in proteins, only those hydrogens located at peptide amide linkages (also referred to as backbone amide hydrogens) have exchange rates in a range that can easily be measured by mass spectrometry. Since every residue except proline has one such backbone amide hydrogen, this class of hydrogens provides a series of discrete sensors extending the entire length of the polypeptide chain. Nearly all backbone amide hydrogens in folded proteins are hydrogen bonded, either intramolecularly to another part of the protein or to water. A large reduction in amide hydrogen exchange rates occurs in folded proteins, primarily due to restricted access of solvent to the interior of the protein and to intraprotein hydrogen bonding. In folded proteins, some backbone amide hydrogens exchange quickly while others exchange only after months. The rates of the most slowly exchanging amide hydrogens may be reduced by as much as 10^8 compared to their rates in unfolded forms of the same protein [3].

At physiological pH, base-catalyzed exchange is the dominant mechanism for hydrogen exchange. Base-catalyzed isotope exchange occurs only when a hydrogen bond is severed in the presence of the catalyst (hydroxide) and the source of the new hydrogen (water). Abstraction of the amide proton by hydroxide is followed by reprotonation/deuteration of the amide nitrogen with a proton/deuteron from the solvent. For slowly exchanging hydrogens, the exchange requires structural fluctuations that free these hydrogens from intramolecular hydrogen bonding and provide access to the aqueous solvent. These structural changes may be of low amplitude, involving only a few atoms, or the structural changes may involve movements of large segments of the polypeptide backbone, or even complete unfolding of the protein. If either hydrogen bonding or solvent accessibility is altered during protein interactions, the hydrogen exchange rate(s) will be changed.

3.3   GENERAL HX-MS METHOD

The methods used for HX-MS analysis have been reviewed previously [7, 12, 13]. Proteins are placed in physiological buffers at room temperature such that they are in their native state. Deuterium is added in a significant excess (usually >15–20-fold by volume) to drive the labeling reaction completely in one direction. After a set period of exchange, usually lasting from 1 second to as long as 24 hours, the exchange reaction is quenched by lowering the pH to 2.5. Because HX is both acid and base catalyzed, there is a minimum exchange rate at around pH 2.5 [3, 14, 15]. By reducing the pH from near neutral to 2.5, the exchange rate decreases by a factor of approximately 10^4. To further reduce the exchange rate, the temperature is lowered from room temperature to 0 °C, reducing the exchange rate by another order of magnitude. Under these quench conditions, the deuterium label would revert to hydrogen, in a process known as back-exchange, with a half-life between 30 and 60 minutes (dependent on sequence) if the protein were exposed to 100% H_2O. In order to introduce the labeled protein sample into the mass spectrometer with electrospray ionization (alternatively, MALDI can be used; see Reference [11]), an HPLC (high-performance liquid chromatography) step is required in which protiated (e.g., H_2O) solvents are used. Back-exchange from deuterium to hydrogen occurs during the HPLC step. To minimize the level of the back-exchange, the chromatography is carried out under quench conditions (pH 2.5, 0 °C). Control samples can be prepared that allow adjustment for back-exchange [12, 16].

The use of HX-MS to investigate protein interactions is illustrated schematically in Figure 3.1. These techniques can be applied to any ligand and are not necessarily restricted to protein–protein interactions. Generally, HX into the intact protein(s) is measured first (Figure 3.1a). The proteins are independently exposed to D_2O at near neutral pH for 10–15 selected periods

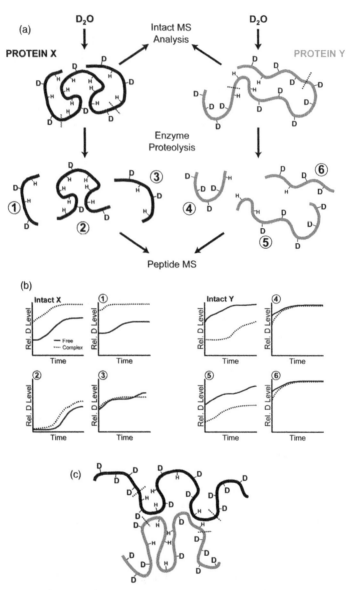

Figure 3.1 Example of applying hydrogen exchange mass spectrometry to the study of protein interactions. (a) Two proteins, X (black line, left) and Y (gray line, right) are independently incubated in D₂O for distinct periods of time. The mass of each protein can be determined without digestion (top center) or the labeled proteins can be digested into small peptides for analysis (middle). The digestion of protein X produces three fragments (numbered 1–3) and the digestion of protein Y produces three fragments (numbered 4–6), which are then subjected to mass analysis (bottom). (b) Hypothetical deuterium uptake curves for the intact proteins in panel (a) (upper left in each group) and the peptic peptides from each protein. The free protein data are shown as solid lines and the complex as dotted lines. (c) A hypothetical complex between proteins X and Y, shown after incubation in D₂O.

of time (usually ranging from 1 second to 24 hours). The deuterium exchange reaction is quenched by reducing the pH to 2.5 and the temperature to 0 °C. The quenched proteins are separately directed into an HPLC to remove the buffer salts and the eluate is sent into an electrospray mass spectrometer for mass analysis (see References [7, 12, 13] for details). Deuterium incorporation over time is plotted for each intact protein (Figure 3.1b, top left panels). These deuterium uptake curves in themselves provide information about how many amide positions exchanged rapidly, slowly, or not at all during the time course. However, the increase in the mass of each protein after hydrogen exchange only provides an overview of the extent of change, not the location of the changes.

3.3.1   Location Information Provided by HX-MS

To localize the changes in deuterium uptake that are observed in intact protein(s), the protein(s) can be digested with a proteolytic enzyme after the exchange reaction has been quenched. A number of groups have effectively used similar methodology in experiments to probe protein–ligand and protein–protein interactions [17–23]. For the example proteins illustrated here (Figure 3.1), pepsin digestion of the HX quenched samples generated three peptides per protein. It is important to note that pepsin digestion is carried out after hydrogen exchange is completed and quenched. Hence, the structural information captured by deuterium incorporation is present in the peptides. The mass of the peptides is determined in the same way as the mass of the intact protein and the deuterium uptake is plotted as a function of time (Figure 3.1b). As was the case for analysis of intact proteins, the deuterium uptake curves provide information about how many amide positions exchanged rapidly, slowly, or not at all during the time course. Because the amino acid sequence of each peptic peptide can be determined from its mass and from tandem MS experiments, the deuterium uptake information can be mapped onto specific regions of the protein. If structural information is already available for the protein being analyzed, the hydrogen exchange information can then be correlated with the existing structural data. This type of information gives clues about the local structural environment of the protein. HX-MS data can also be useful in the absence of any structural data. Although it is not possible to explicitly determine secondary or tertiary structure with HX-MS, regions exhibiting limited deuterium uptake indicate the presence of structure. When combined with other information (such as secondary structure predictions, limited proteolysis information, or homology modeling), rough conclusions can be drawn about the structure of proteins for which there is no high-resolution structural information [24].

3.3.2   Revealing Interactions by Comparison

HX-MS data is extremely useful when different states of the same protein are compared (for examples, see References [23–33]). If the two proteins in Figure 3.1a were to interact with each other (as illustrated in Figure 3.1c), then changes in the HX result. By obtaining baseline data for the proteins when they are alone and comparing it to the data for the same proteins in the complex, regions that were involved in the interaction between the two proteins could be determined. Figure 3.1b illustrates some hypothetical data for proteins X and Y. Comparison of the deuterium uptake curves for protein X when free or in complex with protein Y indicates that more deuterium was incorporated in protein X in the complex. Such results imply that conformational changes occurred that exposed more amide hydrogens to deuterium. The location of the changes is then determined by comparing the deuterium uptake curves for the peptides. For this hypothetical data, peptide 1 experienced significant increases in deuterium uptake while exchange into peptides 2 and 3 was not significantly altered. One concludes, therefore, that the region of protein X that was most altered by binding to protein Y must have included the amino acids located in peptide 1. Similar comparisons can be made for all the peptides of both proteins.

Recall that this type of analysis can be performed for any type of ligand. For example, protein X could be probed in the presence/absence of another protein or protein complex, a binding peptide, a small molecule, a lipid, a polysaccharide, or a nucleic acid. Although only two proteins are shown in Figure 3.1, these experiments can just as well be done with more than two interacting proteins, offering the potential for studying the binding interactions and conformational changes in large, multimeric protein complexes. As long as the interactions between the partners causes a change to the amount and/or location of hydrogen exchange, it will be detectable. Caution must be used, however, because not all interactions may cause changes, as described in the following section.

3.4   INTERACTIONS OF PROTEINS

Protein interactions are generally of two types: electrostatic or hydrophobic [34]. The interfaces of the interacting partners may be with the backbone of the polypeptide chain, with the side chains, or a combination of both. It is important to realize that not all interactions will cause changes to the backbone amide hydrogen exchange rates that are being measured with HX-MS. Consider the case in which a small-molecule drug interacts with the functional groups of the side chains in a binding pocket. If the interactions are exclusively with the atoms of the side chains, this may not necessarily cause a change in the backbone amide hydrogen exchange rates of the residues involved. This is particularly true when the backbone amide hydrogens are involved in stable secondary structures such as alpha helices. In one case [28], a complex of two proteins is formed in which one protein (UBC9) interacts with its partner (SUMO1) via an alpha helix. HX-MS measurements showed no detectable change in the hydrogen exchange in UBC9 although NMR chemical shift information clearly showed that the first alpha helix of UBC9 was involved in the binding interface. In contrast, the binding partner SUMO1 had significant changes to deuterium uptake in the presence of UBC9 and many of the changes mapped to the NMR-identified binding interface. While changing the charge of one amino acid in the UBC9 binding-interface alpha helix was sufficient to abolish binding to SUMO1 (as measured by HX-MS of the SUMO1 protein), no alterations in UBC9 HX were detected.

As long as there is a change in HX in some part of the protein, the interactions can be investigated. It is important, however, to distinguish conformationally induced changes from changes caused by occlusion of exposed amide hydrogens. The latter, occlusion, is conceptionally the easiest to understand. Some regions (or peptides) from proteins X and Y in the example in Figure 3.1 may have altered exchange rates in a complex either because the domain or the region is blocked from solvent by the other protein, or as a result of changes that led to a tighter or looser conformation that restricts HX. If two molecules interact with each other, the surface that is involved with the interaction will exclude solvent, and the ability of D_2O to reach the region will be reduced. However, such a phenomenon is also a function of the dissociation constant of the interaction. If the interaction is weak, the partners may spend a significant amount of time apart. Under these circumstances, D_2O is able to exchange at the interface with the same efficiency as it does when the two partners are probed in isolation. By measuring the exchange at short time periods (1–10 seconds of labeling), the amide hydrogens that are occluded can be separated from those that are not involved in the exact binding interface [35, 36].

Conformational changes arising from interactions are a more difficult problem. In a number of cases (see later discussion), allosteric changes and alterations to protein stability and the folded structure can cause changes in HX in regions quite distant from the interaction surface [11, 37]. A pulse-chase strategy [38] can be used to differentiate these conformationally induced changes from local changes due to solvent occlusion. The basic approach is to initiate exchange in D_2O with only one partner (the protein of interest) for a set period of time, say, 30 minutes. The interacting partner is added, the D_2O is replaced with H_2O, and the exchange is allowed to proceed for the same period of time. At the end of the exchange period, the sample is subjected to mass analysis. If the interacting partner is not present, the protein is expected to be completely unlabeled after such an experiment because all the forward labeling with D_2O would be reversed by the equivalent amount of time for exchange in H_2O. If the interaction that began when the interacting partner was added at the D_2O-H_2O switchover caused a change in the accessibility of some sites, then some quantity of deuterium would remain at the end of the experiment.

The dissociation constants for interactions can be obtained with HX-MS simply by monitoring the change in HX as ligand concentration is changed (see also Reference [13]). In the PLIMSTEX method [39, 40], for example, binding constants can be extracted for multiple binding events by monitoring the change in HX behavior upon binding. These methods offer advantages over fluorescence-based binding assays, which often require either the replacement of native amino acids with reporting tryptophans in the vicinity of the binding site or the use of fluorescent dyes (e.g., ANS), which may disrupt the native structure. HX-MS is done in solution and does not require that one of the binding partners be immobilized as in surface plasmon resonance and similar methods.

3.5 EXAMPLES

3.5.1 Conformational Changes of Proteins During Binding

Many, but not all, proteins undergo some type of change during interactions, which is detectable by HX-MS. Proteins that have substantial changes in conformation during interactions are especially easy to probe with HX-MS. It is becoming increasingly obvious that a number of proteins exist in an unfolded state, only folding when they make contact with their binding partner(s) [41]. One such recent example is that of the carboxy-terminal domain of HIF-1 alpha [42], which exists in a mostly unfolded and disordered state until it finds its binding partner. Upon interaction, three new helices are formed in the HIF protein and there are significant hydrophobic interactions with the target protein and HIF. The HX-MS of HIF-1 alpha protein when alone should be significantly different from that of the bound form. Similar experiments could be performed on any system. In the following examples, the use of HX-MS to probe protein interactions will be described.

3.5.2 Protein–Protein Interactions

The human heat shock protein 70 (Hsp70) is a chaperone that assists in protein folding in the cytosol (reviewed in Reference [43]). In a recent set of experiments, we have used HX-MS to investigate Hsp70 interactions with target

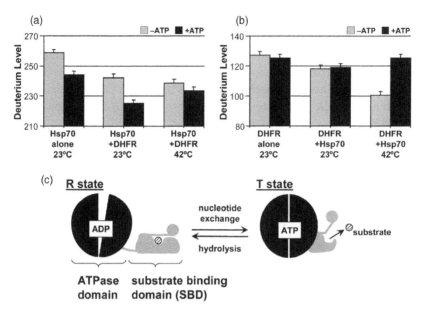

Figure 3.2 Example of using HX-MS to analyze protein–protein interactions. The proteins Hsp70 (a) and DHFR (b) were investigated in the presence and absence of ATP. See the text for the details of each experiment. (c) Model for the conformation of the two states of Hsp70 [44].

proteins, including proteins unfolded by heat. The experimental protocol illustrated in Figure 1 was followed for purified Hsp70 alone, the target protein dihydrofolate reductase (DHFR), and a 1:1 molar mixture of the two proteins. The proteins were allowed to interact for 30 minutes and then labeled with deuterium for 1 hour. The deuterium levels of Hsp70 (Figure 3.2a) and DHFR (Figure 3.2b) were measured. Comparison of the deuterium uptake for Hsp70 with and without DHFR indicates that Hsp70 conformation is altered by the presence of DHFR such that the deuterium level after 1 hour drops from ~260 to ~245 (Figure 3.2a, left, middle). However, the differences in DHFR deuterium level are much less striking, dropping by only about 8–10 deuterons (Figure 3.2b, left and middle).

As Hsp70 is an ATPase, it was also incubated with and without ATP for the HX-MS experiments. Comparing the deuterium uptake in the presence of ATP to the uptake in the absence of ATP for Hsp70 at room temperature (Figure 3.2a, left), it is clear that something significant also happens to Hsp70 in the presence of ATP that inhibits the exchange. Addition of DHFR to Hsp70 + ATP (Figure 3.2a, middle) causes a further reduction in deuterium uptake that is similar in magnitude to the reduction in deuterium levels when ATP was incubated with Hsp70 without DHFR present. These results indicate that the decrease in deuterium uptake caused by DHFR is not the same as that caused by ATP. The model for the Hsp70 function (Figure 3.2c) supports this idea in that the ATPase domain is separate from the substrate binding domain [44]. According to the model, ATP is expected to reduce the solvent accessibility of the ATPase domain by altering its conformation and converting it from an open state (R-state) to a closed state (T-state). In contrast, ATP did not change the deuterium uptake of DHFR at all (Figure 3.2b, left and middle), consistent with the absence of an interaction between ATP and DHFR.

Hsp70 plays a role in helping proteins refold after heat stress when they may be prone to unfolding and aggregation. HX-MS analysis was repeated for DHFR that had been heat-treated for 30 minutes at 42 °C, mixed with Hsp70 at 23 °C, allowed to interact for 20 minutes, and then labeled with deuterium for 1 hour. The experiment was carried out in the presence and absence of ATP. Because ATP is required for the refolding activity of Hsp70, it was expected that the deuterium uptake of heat-treated DHFR/Hsp70(+ATP) would be different from DHFR/Hsp70(−ATP). Figure 3.2b shows that this was indeed the case. Heat-treated DHFR in the presence of Hsp70 alone only took up ~100 deuterons after 1 hour of labeling. This was in stark contrast to the identical experiment without heat treatment in which it was able to incorporate ~130 deuterons. The difference is attributed to aggregation of DHFR caused by unfolding at 42 °C. Hsp70 in the presence of ATP was able to reverse the aggregation and return the DHFR to a state capable of incorporating nearly the same amount of deuterium as it did at 23 °C (Figure 3.2b, right). Such results demonstrate that Hsp70 is indeed a molecular chaperone that can rescue aggregated proteins and that it requires ATP to do so. Furthermore, these experiments illustrate several ways in which protein interactions can be probed with HX-MS. The ATP (small-molecule) interaction with Hsp70 caused changes to deuterium incorporation, whereas it had no effect on DHFR. Pepsin digestion experiments could be used to locate where the conformation of Hsp70 was altered in the presence of ATP (see also another small-molecule example later). In addition, the protein–protein interactions between Hsp70 and DHFR were also investigated and HX-MS revealed that the interaction between the two had a more dramatic effect on Hsp70 than it did on DHFR. Again, pepsin digestion experiments could reveal where the changes occurred and could help to identify the binding interface (as described earlier).

3.5.3 Protein–Peptide Interactions

The interaction between proteins and peptides (8–25 amino acids) can be investigated by HX-MS provided some change occurs in the protein. Due to dissociation constant constraints, a large excess of peptide is often required to ensure that >75% of protein molecules are bound during labeling (see

References [45, 46]). Much of the peptide, therefore, is not bound at any given time and is therefore almost completely deuterated upon final analysis. We have shown that binding of the 12 amino acid HIV Nef peptide to the Hck SH3 domain (K_d = 90 μM) altered the dynamics of the Hck SH3 domain such that the K_d could be determined and the regions of Hck SH3 that had altered dynamics upon binding were identified [45]. In the case of the Hck SH3 domain, slowed dynamics were observed on the side of the protein opposite to the peptide binding interface, indicating that peptide binding slows the overall dynamics or breathing motions of the domain. Similar results were obtained for the Lck SH3 domain when bound to a peptide (10 μM K_d) [47] and for the Abl SH3 domain when bound to the BP1 peptide [48] (data not shown). For the Hck SH3 domain, increasing the local concentration of the peptide by tethering it on to the protein was found to be an effective method for making the differences in HX-MS more dramatic in the bound versus free forms of the protein [46, 49]. Increasing the affinity of peptides for their binding partners by mutation (See Reference [46]) is also a way to enhance the effects of binding on the changes to HX-MS.

3.5.4   Protein–Small Molecule Interactions

The interactions between DHFR and a small-molecule inhibitor illustrate the application of HX-MS to probe interactions between proteins and small molecules. DHFR is an enzyme that converts dihydrofolate into tetrahydrofolate and is thus required for the synthesis of purines. Methotrexate (MTX) is a potent inhibitor of DHFR and has been used in the treatment of breast cancer and rheumatic diseases [50, 51]. HX-MS was used to determine how MTX binding alters the dynamics of mouse DHFR. Figure 3.3 provides a representative illustration of the type of information that HX-MS provides in this kind of experiment. HX-MS of DHFR peptic peptides was performed as depicted in Figure 3.1. Comparison of HX-MS data for DHFR in the presence and absence of MTX revealed the regions of DHFR that were altered by MTX binding. Some regions of DHFR experienced no change in deuterium incorporation in the presence of MTX (Figure 3.3a), while others showed moderate (Figure 3.3b) to large (Figure 3.3c) changes in deuterium incorporation. Mapping the HX-MS data onto the structure of DHFR with MTX bound indicates which parts of the protein experienced changes in deuterium incorporation in the presence of MTX. As shown in Figure 3.3d, some regions did not experience any changes at all, including the small flap portion from residues 56 to 65 that is directly located above the binding pocket. The lower half of DHFR experienced a general decrease in protein dynamics, as indicated by the reduction of deuterium incorporation throughout the peptic peptides in this area. While the diagram suggests that the entire lower half of DHFR

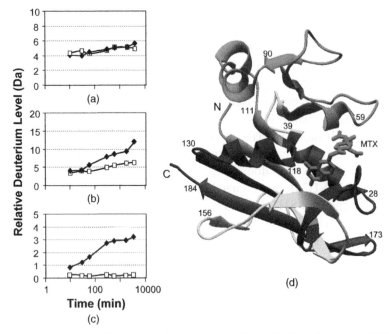

Figure 3.3   Protein–small molecule interactions probed with HX-MS. (a–c) DHFR was incubated with (open symbol) or without (solid symbol) methotrexate and deuterium incorporation measured for various times in D_2O. The error of each data point is ± 0.5 Da. (d) Summary of regions of DHFR that experience changes in deuterium uptake in the presence of methotrexate. No data were obtained for regions that are colored gray. DHFR coordinates were from PDB file 1DHF.

is affected, such illustrations can be misleading. The illustration implies that the effects extend over the entire length of the peptides, while it is usually the case that the affected regions are much smaller. For example, in Figure 3.3b, there is a reduction by approximately 4–5 deuterons in the peptide covering DHFR residues 113–134, which contains 20 backbone amide hydrogens. Because the location of the deuterium within the peptic peptide cannot be determined [11], it is not possible to identify the specific amide hydrogens that are affected within this region. The use of overlapping peptides and more efficient digestion can reduce this limitation [16, 52] by more narrowly localizing the regions that are affected.

The results presented in Figure 3.3 illustrate several important points about probing protein–small molecule interactions with HX-MS. First, small molecules can induce significant changes in regions distant from their binding sites. The greater this effect, the more difficult it becomes to identify the binding interface as the large changes in HX caused by the conformational changes will overwhelm small changes at the binding interface. Second, the binding interface encompasses small regions that are not adjacent in the primary structure; thus, these small regions will be spread over different peptides. Sometimes it is difficult to localize changes of a few amide hydrogens when they are spread out in this fashion. Third, while MTX interacted mostly with side chain atoms, there was a decrease in HX in secondary structural elements (i.e., the alpha helix in Figure 3.3c, residues 33–38), likely as a result of changes to the dynamics or breathing motions of the protein. Fourth, incubation of DHFR with small molecules that do not bind (such as ATP illustrated in Figure 3.2) did not cause changes in HX. By monitoring the changes associated at some obvious point, say, in peptides 33–38 after 2 hours in D_2O (Figure 3.3c), small molecules can be screened for binding to DHFR. Similar binding to that of MTX should induce similar changes in HX at this one time point. Thus, limited HX-MS can be used as a relatively simple assay for small-molecule binding.

3.6   CONCLUSIONS

The use of hydrogen exchange mass spectrometry for probing protein interactions has many advantages. There are also some limitations, as have been pointed out in the few examples given. As with any method, HX-MS has its own unique contributions to make to the study of proteins. When combined with other approaches, such as site-directed mutagenesis, the value of HX-MS is only magnified. It is hoped that many more protein interactions will be uncovered with the help of HX-MS.

ACKNOWLEDGMENTS

We are pleased to acknowledge T. Sibray for assistance in protein expression and purification. This work was supported by funding from the NIH (R01-GM070590 and P20-RR016480).

REFERENCES

1. Hvidt, A., and Nielsen, S. O. (1966). Hydrogen exchange in proteins. *Adv. Protein Chem.* **21**: 287–385.
2. Woodward, C., Simon, I., and Tüchsen, E. (1982). Hydrogen exchange and the dynamic structure of proteins. *Mol. Cell. Biochem.* **48**: 135–160.
3. Englander, S. W., and Kallenbach, N. R. (1984). Hydrogen exchange and structural dynamics of proteins and nucleic acids. *Q. Rev. Biophys.* **16**: 521–655.

4. Kim, K.-S., and Woodward, C. (1993). Protein internal flexibility and global stability: effect of urea on hydrogen exchange rates of bovine pancreatic trypsin inhibitor. *Biochemistry* **32**: 9609–9613.

5. Miller, D. W., and Dill, K. A. (1995). A statistical mechanical model for hydrogen exchange in globular proteins. *Protein Sci.* **4**: 1860–1873.

6. Engen, J. R., and Smith, D. L. (2001). Investigating protein structure and dynamics by hydrogen exchange MS. *Anal. Chem.* **73**: 256A–265A.

7. Engen, J. R., and Smith, D. L. (2000). Investigating the higher order structure of proteins: hydrogen exchange, proteolytic fragmentation & mass spectrometry. *Meth. Mol. Biol.* **146**: 95–112.

8. Smith, D. L., Deng, Y., and Zhang, Z. (1997). Probing the non-covalent structure of proteins by amide hydrogen exchange and mass spectrometry. *J. Mass. Spectrom.* **32**: 135–146.

9. Kaltashov, I. A., and Eyles, S. J. (2002). Studies of biomolecular conformations and conformational dynamics by mass spectrometry. *Mass Spectrom. Rev.* **21**: 37–71.

10. Hoofnagle, A. N., Resing, K. A., and Ahn, N. G. (2003). Protein analysis by hydrogen exchange mass spectrometry. *Annu. Rev. Biophys. Biomol. Struct.* **32**: 1–25.

11. Wales, T. E., and Engen, J. R. (2006). Hydrogen exchange mass spectrometry for the analysis of protein dynamics. *Mass Spectrom. Rev.* **25**: 158–170.

12. Hoofnagle, A. N., Resing, K. A., and Ahn, N. G. (2004). Practical methods for deuterium exchange/mass spectrometry. *Meth. Mol. Biol.* **250**: 283–298.

13. Kaveti, S., and Engen, J. R. (2006). Protein interactions probed with mass spectrometry. In *Bioinformatics and Drug Discovery* (R. S. Larson, Ed.), Vol. 316, pp. 179–197, Humana Press, Totowa, NJ.

14. Molday, R. S., Englander, S. W., and Kallen, R. G. (1972). Primary structure effects on peptide group hydrogen exchange. *Biochemistry* **11**: 150–158.

15. Bai, Y., Milne, J. S., Mayne, L., and Englander, S. W. (1993). Primary structure effects on peptide group hydrogen exchange. *Proteins: Struct. Funct. Genet.* **17**: 75–86.

16. Zhang, Z., and Smith, D. L. (1993). Determination of amide hydrogen exchange by mass spectrometry: a new tool for protein structure elucidation. *Protein Sci.* **2**: 522–531.

17. Mandell, J. G., Baerga-Ortiz, A., Akashi, S., Takio, K., and Komives, E. A. (2001). Solvent accessibility of the thrombin–thrombomodulin interface. *J. Mol. Biol.* **306**: 575–589.

18. Maier, C. S., Schimerlik, M. I., and Deinzer, M. L. (1999). Thermal denaturation of *Escherichia coli* thioredoxin studied by hydrogen/deuterium exchange and electrospray ionization mass spectrometry: monitoring a two-state protein unfolding transition. *Biochemistry* **38**: 1136–1143.

19. Engen, J. R., Bradbury, E. M., and Chen, X. (2002). Using stable-isotope-labeled proteins for hydrogen exchange studies in complex mixtures. *Anal. Chem.* **74**: 1680–1686.

20. Chen, J., and Smith, D. L. (2000). Unfolding and disassembly of the chaperonin GroEL o ccurs v ia a t etradecameric i ntermediate wi th a f olded e quatorial domain. *Biochemistry* **39**: 4250–4258.

21. Resing, K. A., and Ahn, N. G. (1998). Deuterium exchange mass spectrometry as a probe of protein kinase activation. Analysis of wild-type and constitutively active mutants of MAP kinase kinase-1. *Biochemistry* **37**: 463–475.

22. Wang, F., Scapin, G., Blanchard, J. S., and Angeletti, R. H. (1998). Substrate binding a nd con formational c hanges o f *Clostridium glutamicum* di aminopimelate dehydrogenase re vealed b y h ydrogen/deuterium e xchange a nd e lectrospray mass spectrometry. *Protein Sci.* **7**: 293–299.

23. Zhang, Y. H., Yan, X., Maier, C. S., Schimerlik, M. I., and Deinzer, M. L. (2002). Conformational analysis of intermediates involved in the in vitro folding pathways of recombinant human macrophage colony stimulating factor beta by s ulfhydryl g roup t rapping a nd h ydrogen/deuterium pu lsed lab eling. *Biochemistry* **41**: 15495–15504.

24. Hamuro, Y., B urns, L ., Ca naves, J ., H offman, R ., T aylor, S ., a nd W oods, V. (2002). Domain organization of D-AKAP2 revealed by enhanced deuterium exchange–mass spectrometry (DXMS). *J. Mol. Biol.* **321**: 703–714.

25. Hamuro, Y., Wong, L., Shaffer, J., Kim, J. S., Stranz, D. D., Jennings, P. A., Woods, V. L. Jr., and Adams, J. A. (2002). Phosphorylation driven motions in the COOH-terminal Src kinase, CSK, revealed through enhanced hydrogen–deuterium e xchange a nd m ass sp ectrometry (DXMS). *J. Mol. Biol.* **323**: 871–881.

26. Anand, G. S., Hughes, C. A., Jones, J. M., Taylor, S. S., and Komives, E. A. (2002). Amide H/2H exchange reveals communication between the cAMP and catalytic subunit-binding sites in the R(I)alpha subunit of protein kinase A. *J. Mol. Biol.* **323**: 377–386.

27. Resing, K. A., and Ahn, N. G. (1998). Deuterium exchange mass spectrometry as a probe of protein kinase activation. Analysis of wild-type and constitutively active mutants of MAP kinase kinase-1. *Biochemistry* **37**: 463–475.

28. Engen, J. R. (2003). Analysis of protein complexes with hydrogen exchange and mass spectrometry. *Analyst (London)* **128**: 623–628.

29. Engen, J. R., Smithgall, T. E., Gmeiner, W. H., and Smith, D. L. (1999). Comparison o f S H3 a nd S H2 do main d ynamics w hen e xpressed a lone or i n an SH(3 + 2) construct: the role of protein dynamics in functional regulation. *J. Mol. Biol.* **287**: 645–656.

30. Remigy, H., Jaquinod, M., Petillot, Y., Gagnon, J., Cheng, H., Xia, B., Markley, J. L., Hurley, J. K., Tollin, G., and Forest, E. (1997). Probing the influence of mutation on the stability of a ferredoxin by mass spectrometry. *J. Protein Chem.* **16**: 527–532.

31. Hasan, A., S mith, D. L ., a nd S mith, J. B. (2 002). A lpha-crystallin re gions affected b y a denosine 5 ′-triphosphate i dentified b y h ydrogen–deuterium exchange. *Biochemistry* **41**: 15876–15882.

32. Yan, X., Zhang, H., Watson, J., Schimerlik, M. I., and Deinzer, M. L. (2002). Hydrogen/deuterium exchange and mass spectrometric analysis of a protein containing multiple disulfide bonds: solution structure of recombinant macrophage colony stimulating factor-beta (rhM-CSFbeta). *Protein Sci.* **11**: 2113–2124.

33. Kim, M. Y., Maier, C. S., Reed, D. J., and Deinzer, M. L. (2002). Conformational changes in chemically modified *Escherichia coli* thioredoxin monitored by H/D exchange and electrospray ionization mass spectrometry. *Protein Sci.* **11**: 1320–1329.

34. Creighton, T. E. (1993). *Proteins: Structures and Molecular Properties*, 2nd. ed., Trans., W.H. Freeman and Company, New York.

35. Dharmasiri, K., and Smith, D. L. (1996). Mass spectrometric determination of isotopic exchange rates of amide hydrogens located on the surfaces of proteins. *Anal. Chem.* **68**: 2340–2344.

36. Mandell, J. G., Baerga-Ortiz, A., Akashi, S., Takio, K., and Komives, E. A. (2001). Solvent accessibility of the thrombin–thrombomodulin interface. *J. Mol. Biol.* **306**: 575–589.

37. Mayne, L., Paterson, Y., Cerasoli, D., and Englander, S. W. (1992). Effect of antibody binding on protein motions studied by hydrogen-exchange labeling and two-dimensional NMR. *Biochemistry* **31**: 10678–10685.

38. Garcia, R. A., Pantazatos, D., and Villarreal, F. J. (2004). Hydrogen/deuterium exchange mass spectrometry for investigating protein–ligand interactions. *Assay Drug Dev. Technol.* **2**: 81–91.

39. Zhu, M. M., Rempel, D. L., Du, Z., and Gross, M. L. (2003). Quantification of protein–ligand interactions by mass spectrometry, titration, and H/D exchange: PLIMSTEX. *J. Am. Chem. Soc.* **125**: 5252–5253.

40. Zhu, M. M., Rempel, D. L., and Gross, M. L. (2004). Modeling data from titration, amide H/D exchange, and mass spectrometry to obtain protein–ligand binding constants. *J. Am. Soc. Mass Spectrom.* **15**: 388–397.

41. Dyson, H. J., and Wright, P. E. (2004). Unfolded proteins and protein folding studied by NMR. *Chem. Rev.* **104**: 3607–3622.

42. Dames, S. A., Martinez-Yamout, M., De Guzman, R. N., Dyson, H. J., and Wright, P. E. (2002). Structural basis for Hif-1 alpha/CBP recognition in the cellular hypoxic response. *Proc. Natl. Acad. Sci. U. S. A.* **99**: 5271–5276.

43. Hartl, F. U., and Martin, J. (1995). Molecular chaperones in cellular protein folding. *Curr. Opin. Struct. Biol.* **5**: 92–102.

44. Rudiger, S., Buchberger, A., and Bukau, B. (1997). Interaction of Hsp70 chaperones with substrates. *Nat. Struct. Biol.* **4**: 342–349.

45. Engen, J. R., Smithgall, T. E., Gmeiner, W. H., and Smith, D. L. (1997). Identification and localization of slow, natural, cooperative unfolding in the hematopoietic cell kinase SH3 domain by amide hydrogen exchange and mass spectrometry. *Biochemistry* **36**: 14384–14391.

46. Hochrein, J. M., Lerner, E. C., Schiavone, A. P., Smithgall, T. E., and Engen, J. R. (2006). An examination of dynamics crosstalk between SH2 and SH3

domains by hydrogen/deuterium exchange and mass spectrometry. *Protein Sci.* **15**: 65–73.

47. Weis, D. D., Kjellen, P., Sefton, B. M., and Engen, J. R. (2006). Altered dynamics in Lck SH3 upon binding to the LBD1 domain of *Herpesvirus saimiri* tip. *Protein Sci.* **15**: 2411–2422.

48. Chen, S., Brier, S., Smithgall, T. E., and Engen, J. R. (2007). The Abl SH2-kinase linker naturally adopts a conformation competent for SH3 domain binding. *Protein Sci.* **16**: 572–581.

49. Gmeiner, W. H., Xu, I., Horita, D. A., Smithgall, T. E., Engen, J. R., Smith, D. L., and Byrd, R. A. (2001). Intramolecular binding of a proximal PPII helix to an SH3 domain in the fusion protein SH3Hck: PPIIhGAP. *Cell. Biochem. Biophys.* **35**: 115–126.

50. Schweitzer, B. I., Dicker, A. P., and Bertino, J. R. (1990). Dihydrofolate reductase as a therapeutic target. *FASEB J.* **4**: 2441–2452.

51. Chan, D. C., and Anderson, A. C. (2006). Towards species-specific antifolates. *Curr. Med. Chem.* **13**: 377–398.

52. Cravello, L., Lascoux, D., and Forest, E. (2003). Use of different proteases working in acidic conditions to improve sequence coverage and resolution in hydrogen/deuterium exchange of large proteins. *Rapid Commun. Mass Spectrom.* **17**: 2387–2393.

4

LIMITED PROTEOLYSIS MASS SPECTROMETRY OF PROTEIN COMPLEXES

MARIA MONTI AND PIERO PUCCI

Dipartimento di Chimica Organica e Biochimica, Università di Napoli "Federico II," Napoli, Italy

4.1 I ntroduction
4.2 Limited Proteolysis Analysis
4.3 E xperimental Design
4.4 Probing Protein–Protein Interactions
4.5 Probing Protein–Nucleic Acid Interactions
4.6 Probing Protein–Ligand Interactions
4.7 Probing Amyloid Fibril Core
4.8 Conc lusions
    References

4.1 INTRODUCTION

A critical step toward the understanding of interactions occurring in protein complexes consists in the structural description of the contact regions within the complex. These studies are usually accomplished either in the solid state by X-ray cr ystallography o r i n s olution b y m ultidimensional N MR [1, 2].

Mass Spectrometry of Protein Interactions Edited by Kevin M. Downard
Copyright © 2007 John Wiley & Sons, Inc.

However, difficulties in the crystallization of the complex, limited solubility of the macromolecules in aqueous solvent, or the requirement of large amounts of material might severely impair these investigations. Thus, there has been an increasing demand for alternative approaches able to provide experimental data to usefully supplement X-ray and NMR studies in the description of the topology of protein complexes.

The main structural feature of interacting proteins in protein complexes is that the interface regions are usually accessible to the solvent in the isolated molecules, whereas they become protected following the formation of the complex. Moreover, it is conceivable that, following interaction, proteins undergo conformational changes that lead the complex to assume its proper biologically active structure. Based on these observations, a series of experimental approaches, which combine classical protein chemistry methodologies with mass spectrometric procedures, have been designed to probe the topology of interacting regions in protein complexes. Among these, limited proteolysis experiments carried out on both the isolated proteins and the complex using different proteases followed by mass spectrometric analysis of the fragments generated can be employed to identify proteolytically accessible amino acid residues. Since the contact regions of the two interacting proteins in the complex are protected against proteolysis, differential peptide maps are obtained from which the definition of the interface zones as well as regions affected by conformational changes can be inferred.

This chapter focuses on limited proteolysis as a tool for the investigation of protein complexes using some exemplary studies to highlight its utility, as well as pointing out the potential pitfalls.

4.2 LIMITED PROTEOLYSIS ANALYSIS

Limited proteolysis of native proteins had largely been used by biochemists in the past in order to retrieve structural information on the three-dimensional structure of the protein. The premise underpinning such studies is the sequence/structure paradigm of limited proteolysis: higher order structure and not primary sequence is the main determinant of the site of preferential hydrolysis. When the digestion reaction is "limited" by controlling different experimental parameters, protease activity is preferentially addressed toward a few, specific peptide bonds. Conformational parameters such as accessibility, segmental mobility, and solvent exposition correlate quite well with limited proteolytic sites. However, as pointed out by Fontana et al. [3], these parameters are themselves highly correlated and are not enough to rationalize the limited proteolysis phenomenon. What remains unchallenged are the local dynamic properties of the polypeptide chain, which are needed for local unfolding to take place to accommodate peptide bonds within the protease active site [4].

Limited proteolysis was first employed to identify and to isolate individual domains from multidomain proteins [5, 6]. Although not exclusively true, protein domains are structural elements often endowed with their own biological functions and connected to each other by relatively unstructured polypeptide linkers. Limited proteolysis and the subsequent identification of protease-resistant fragments offer an experimental route for the determination of exact boundaries of the domains. The basic premise of this approach is that proteolysis of multidomain proteins occurs more readily in the less structured linker regions between the more densely packed domains than within the domains themselves. A classical example of these experiments was the cleavage of RNase A by subtilisin at the 20–21 peptide bond, producing the so-called ribonuclease S made up of a non covalent complex between the S-peptide (1–20) and the remaining protein [7].

Although developed several years ago, this approach still constitutes a powerful tool in the investigation of multidomain proteins as demonstrated by the recent study on the p62 subunit of human transcription/repair factor TFIIH [8]. Limited proteolysis experiments coupled to mass spectrometry identified structural domains within the conserved N-terminal part of the molecule. Biophysical characterization by fluorescence studies and NMR analysis indicated that these domains were at least partially folded and thus may correspond to structural entities. The approach used in this study is general and can be straightforwardly applied to other multidomain proteins and/or multiprotein assemblies.

Besides domain separation, limited proteolysis has been used extensively as a simple method to gain insight into the general fold of proteins of unknown structure and to complement modeling studies. In the early 1990s a number of papers appeared concerning the application of the limited proteolysis approach to a more detailed probing of protein surface [9–12]. Moreover, this approach resulted in a powerful tool for the investigation of the interacting region between protein antigen and the specific antibody. This strategy, known as "epitope mapping," relies on the idea that the antigen region involved in the interaction with the antibody (i.e., the epitope recognized by the antibody variable region) is shielded from the proteases' activities when the antigen–antibody complex is subjected to enzymatic digestion [13–17].

In these applications, most proteolytic experiments were monitored by SDS-PAGE, a simple technique available to most laboratories. Protein portions released during enzymatic hydrolysis were fractionated on the gel and the various protein regions identified by N-terminal sequencing by Edman degradation. However, gel electrophoresis of proteinase digests will rarely yield the precise site of limited proteolysis unless the protease is particularly narrow in its primary specificity and the amino sequence fortuitously disposed.

More recently, as the concept of "controlled" proteolysis was developed, limited proteolysis experiments were carried out under such conditions to limit protease activity to a single proteolytic event that cleaves the protein molecule into two complementary fragments (complementary proteolysis, Figure 4.1). The peptide bond most susceptible to protease action and hence the accessible amino acid residues can be assigned from the identification of the two complementary peptides released from the intact protein. The native conformation of proteins, in fact, provides some stereochemical barriers to enzymatic attack, leaving the exposed and flexible regions accessible to proteases and preventing the occurrence of proteolytic cleavages within the highly structured core of the protein molecule, or at least slowing their kinetics. Consequently, when these experiments are performed on a time course basis using a panel of proteolytic enzymes with different specificities, the pattern of preferential cleavage sites can provide a topological description of the protein surface.

However, the identification of the cleavage sites within the native protein structure during limited proteolysis experiments conducted on a time course basis needs a rapid, sensitive, and definitive analytical methodology. In this respect, advanced mass spectrometric procedures seem to fulfill these

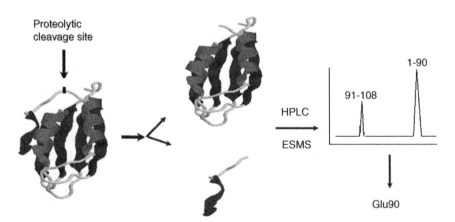

Figure 4.1 Identification of the preferential cleavage sites in the complementary proteolysis mass spectrometry approach. Assignments of the accessible amino acid residues are inferred from the identification of the two complementary peptides released from the intact protein following a single proteolytic event through the determination of their accurate molecular masses.

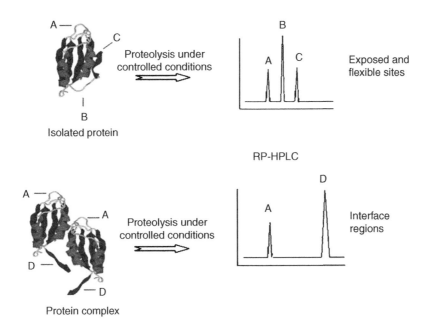

Figure 4.2   Schematic d escription o f t he l imited p roteolysis m ass sp ectrometry st rategy employed in the definition of interface regions in protein complexes.

prerequisites and represent an ideal approach to analyze effectively fragments released enzymatically from the native protein, leading to unambiguous identification of the protease-sensitive sites [18, 19].

This procedure is ideally suited to describe interface regions and to monitor conformational changes occurring in protein structure upon interaction with other molecules (protein complexes). Limited proteolysis experiments carried o ut o n bo th t he i solated pro teins a nd t he co mplex us ing d ifferent proteases f ollowed b y ma ss s pectrometric a nalysis o f t he f ragments g enerated c an b e e mployed t o i dentify p roteolytically a ccessible a mino a cid residues. Since the s urface to pology of t he i nteracting proteins is a ffected by m asking effects a nd/or by c onformational c hanges, d ifferential p eptide maps are obtained from which the definition of the contact surfaces as well as t he re gions i nvolved i n co nformational v ariations ca n b e e xtrapolated (Figure 4.2).

4.3   EXPERIMENTAL DESIGN

Reaction co nditions for l imited pro teolysis ha ve to b e ca refully se lected in order to both ensure maximum stability of the protein and/or protein complex native conformation and to address protease activity toward a few well-defined sites. These conditions include the pH value, the digestion time, and, most importantly, the enzyme to substrate ratio. The pH values related directly to the maintenance of the protein conformation; any unfolding, even partially, of the polypeptide chain, in fact, would result in enzymatic hydrolysis occurring at random, thus leading to an incorrect assessment of the preferred cleavage sites. The proteolysis experiments should be performed at pH values suitable to ensure the stability of the substrate regardless of the optimum pH for enzymatic activity.

The enzyme to substrate ratio is the most important parameter to increase the selectivity of proteases. By varying the amount of proteases in the reaction mixture, it is possible to limit to a great extent their activity, even when broad-specificity proteases had been used. Generally, there are no hard and fast rules governing enzyme:substrate ratios and optimal experimental conditions must be found out by preliminary experimentation. In protein complex analyses, the suitable enzyme to substrate ratio is usually different when the isolated proteins or the protein complex are digested. In this case, proteolysis usually occurs only following a considerable increase in the proteases' concentration. The protein complex, in fact, shows a general lower accessibility than the individual components, because of a tighter compact conformation, compared to the isolated proteins, that decreases the local unfolding events needed for proteolysis. Moreover, the shielding effects exerted by the interacting proteins on each other largely decrease the number of preferential cleavage sites observed in the complex as compared to those found in the isolated molecules.

The enzymatic digestions are often monitored on a time course basis to enhance discrimination between more or less exposed cleavage sites. These sites are referred to as "primary sites" or "secondary sites" merely on a qualitative kinetic basis, the secondary sites being always due to a single proteolytic event, but only in evidence at later stages of the proteolytic process [17, 20]. As the incubation time increases, fragments released from the intact protein are subdigested by the protease, giving rise to smaller peptides; all of these peptides originating from larger fragments and not released from the native proteins do not have to be considered.

Finally, a panel of proteases with different specificities is advisable with the aim to create such conditions where the selectivity of the cleavage is not related to, or limited by, the specificity of the enzyme. The accessibility of the side chain of a particular residue and the local flexibility of the polypeptide chain should eventually be the only factors that define whether or not proteolysis can occur at a particular site. The grouping of the preferential cleavage sites within the same region of the protein regardless of the proteases used is strongly indicative of the consistency of the data and suggestive of the exposure of that particular region.

Identification of the cleavage sites and hence the definition of the exposed residues rely on the characterization of the peptides released from the protein complex. The key role of mass spectrometry in the entire strategy is well illustrated here in that, under limited proteolysis conditions, most of the protein remains undigested and proteolytic fragments are released from the molecule in a very low amount. Mass spectrometric techniques are able to analyze these fragments even when they are present in a minute amount and to measure their mass values with the highest accuracy. The determination of their accurate molecular mass leads to the unambiguous identification of the fragments and hence to the assignment of the cleavage sites.

Two alternative and complementary approaches have been developed so far. First, aliquots of the digestion mixture are withdrawn at different time intervals and fractionated by HPLC. Individual fractions are manually collected and analyzed by electrospray mass spectrometry (ESMS) [21–23]. Besides the determination of the accurate molecular mass of the peptide fragments, a rough estimation of the increasing amount of each peptide released at different times can also be obtained by the HPLC analysis. Alternatively, when the fragments released by enzymatic digest are not amenable to LC separation (low solubility, aggregation problems, etc.), direct MALDI-MS analysis of the proteolytic mixture can also be employed [19, 24, 25]. The extent of proteolytic digestion can be monitored by MALDI analysis on a much shorter time scale using much less protein samples and avoiding freezing of the aliquots.

4.4   PROBING PROTEIN–PROTEIN INTERACTIONS

The potential and limits of the limited proteolysis strategy described previously is well illustrated by the investigation of the topology of the Ca^{2+}–calmodulin (CaM)–melittin ternary complex [26]. The three-dimensional structure of calcium-loaded CaM, determined in solution [27, 28] and in solid state [29], presents two globular domains, each containing two helix–loop–helix motifs, joined by a long α-helix (Figure 4.3). Isolated calcium-bound calmodulin is rapidly digested by proteases with preferential cleavage sites being located within the second half of the long central helix. The two globular domains were more resistant to proteases with only a certain accessibility detected in the C-terminal region. As expected, the unstructured melittin was completely digested by all proteases at nearly all the possible peptide bonds.

However, a completely different proteolytic pattern was observed as a result of the calmodulin–melittin interaction, which highlighted the occurrence of a reciprocal shielding effect exerted by the two interacting molecules. Most of the cleavage sites of calmodulin disappeared and the central region

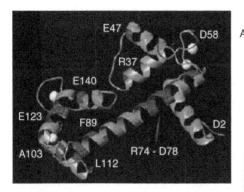

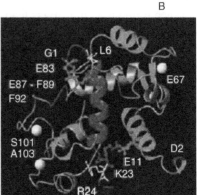

Figure 4.3   (a) Three-dimensional structure of native calmodulin. (b) A view of the three-dimensional st ructure o f t he mo del p roposed f or t he C aM–melittin c omplex sho wing t he reverse orientation of the peptide.

of melittin was fully protected with the complexed peptide being only cleaved at residues located at both ends of the peptide chain.

The a nalysis o f t he Ca M–melittin co mplex us ing t hese i ntegrated a p-proaches s howed t hat, f ollowing t he i nteraction, Ca M u ndergoes a d ramatic conformational change in which the N- and C-terminal domains are brought in close proximity by the disruption of the central helix. The complex then as-sumes a gl obular structure t hat engulfs m ost of t he me littin structure within a h ydrophobic c hannel, l eaving t he se condary s tructure o f t he t wo d omains essentially u nchanged. A lthough t hese e xperimental d ata d id not g ive a ny quantitative structural information, on the basis of these results, a reliable struc-ture of the CaM–melittin complex was modeled using the known structure of the calmodulin–M13 complex and substituting melittin for M13. Surprisingly, the ligand peptide was shown to interact with calmodulin by adopting an oppo-site orientation within the complex as compared to all the peptide substrate ex-amined so far. This finding adds a further dimension to the already remarkable capability of calmodulin in binding different protein substrates, providing the protein with the possibility of regulating an even larger number of enzymatic activities. The "reverse" orientation of melittin within the complex was demon-strated later on by Ikura and co-workers, who solved the NMR structure.

An essentially identical approach was employed to investigate conformational changes occurring within the NS3 protease domain from the hepatitis C virus in the presence of its cofactor Pep4A [30]. Formation of the NS3/Pep4A complex greatly enhances NS3 proteolytic activity, being an absolute requirement for in vivo processing of the viral polyprotein [31]. The affinity of the protease domain for its cofactor peptide is dramatically affected by physicochemical conditions, such as glycerol, salt, and detergent concentration. Limited proteolysis experiments were then performed on the isolated proteins and on the complex under buffer conditions of maximum NS3–Pep4A binding affinity [32]. However, in this particular case, the protease activity at different glycerol concentrations had to be evaluated before performing proteolytic experiments. The decrease of proteolytic activity observed at the higher percentage of glycerol was taken into account to define the suitable enzyme to substrate ratio in each experiment.

The surface topology of isolated NS3 in solution was essentially consistent with the crystal structure of the protein but the N-terminal segment showed an unpredicted high conformational flexibility. Moreover, the region containing the protease active site exhibited an open conformation and a considerable flexibility. At higher glycerol concentration, the protease assumed a more compact structure, showing a decrease in the accessibility of the N-terminal segment and a reduced susceptibility of the active site toward protease action. A similar effect was observed following interaction of NS3 with its cofactor Pep4A. The N-terminal arm was displaced from the protein moiety to accommodate the cofactor peptide, leading the N-terminal segment to adopt again an open and flexible conformation. Again, these results were apparently in contrast with X-ray analysis data depicting a very tight interaction between the NS3 N-terminal region and the cofactor, but were then confirmed by NMR investigations. The conformational changes observed by limited proteolysis were directly correlated with the activation mechanism of the protease exerted by either the cosolvent or the cofactor peptide. The compactness of the N-terminal domain leading to a tighter packing of the substrate binding site, in fact, could be instrumental in a correct alignment of the catalytic triad in the active site.

Perham and collaborators employed the same approach to elucidate the surface topology of the E1 component (pyruvate decarboxylase) of the pyruvate dehydrogenase (PDH) complex from *Bacillus stearothermophilus*. E1p is a heterotetramer consisting of E1α and E1β polypeptide chains associated in an α2β2 stoichiometry [33]. The exposure of the E1p complex to trypsin or chymotrypsin under nondenaturing conditions resulted in digestion of the E1α but not the E1β chain. Furthermore, the whole E1p remained relatively resistant to chymotryptic digestion even for longer times of proteolysis, highlighting a protection and/or compactness effect inferred by the tetrameric complex structure. A qualitative kinetic analysis of the experiments highlighted the occurrence of two independent proteolytic event on the E1α chain, an early cleavage followed by a second slower proteolytic hydrolysis. These data suggested that in the heterotetramer the E1α chains may exist in two different conformations, one more susceptible and one more resistant to protease digestion.

Enzymatic activity assays carried out on the protein complex after limited proteolysis demonstrated that the ability of E1p to decarboxylate pyruvate in the presence of 2,6-dichlorophenolindophenol (DCPIP) was largely increased possibly due to a faster release of the product and/or a higher accessibility of the active site to the DCPIP in the proteolyzed enzyme. However, the overall catalytic activity of PDH containing protease-treated E1p was substantially lower than native PDH due to a lower affinity of proteolyzed E1p for the other PDH protein components. The results achieved by this conformational analysis not only were useful to elucidate the main structure of the E1p polypeptide chains, but allowed the authors to unravel how the E1p conformations could differently affect PDH catalytic activity.

4.5   PROBING PROTEIN–NUCLEIC ACID INTERACTIONS

When the limited proteolysis approach is employed to probe protein–nucleic acid interactions, a further advantage is that the DNA/RNA molecules are resistant to proteases, thus facilitating both the experimental design and the interpretation of the data. However, DNA, RNA, or synthetic oligonucleotides might interfere with the HPLC fractionation of the digestion mixture and have to be removed by selective precipitation before the analysis.

The first example of limited proteolysis experiments applied to the definition of a protein–DNA complex was reported by Brian T. Chait and coworkers who investigated the transcription factor Max, a member of the basic/helix–loop–helix/zipper family of DNA-binding proteins [19]. In this work, the authors had elegantly skipped the interference of DNA on chromatographic peptide fractionation by directly analyzing the digestion product with MALDI-MS. Differently from ESMS, MALDI-MS analyses do not require preventive HPLC separation of peptides. In the absence of DNA and at low ionic strengths, Max is rapidly digested by proteases, suggesting an open and flexible conformation of the protein. At physiological salt levels, a moderate decrease in the digestion rates and the patterns of preferential cleavage sites indicated homodimerization of the protein through a predominantly hydrophobic interface. In the presence of Max-specific DNA, the protein became dramatically protected against proteolysis, exhibiting up to a 100-fold reduction in cleavage rates. A very high degree of protection was detected in the N-terminal and helix–loop–helix regions of the protein, correlating with that expected for a stable dimer bound to DNA at its basic N terminals. Less protection was seen at the C terminal, where proteolytic cleavages occurred correlating to the presence of a leucine zipper. The results also indicated a high affinity of Max for its target DNA even when the leucine zipper was proteolytically removed.

Thyroid transcription factor 1 (TTF-1) is a protein responsible for transcriptional activation of genes expressed in follicular thyroid cells and lung epithelial cells. TTF-1 binds DNA by a homeodomain (HD) of 61 amino acid residues, which specifically recognizes oligonucleotide sequences containing the 5′-CAAG-3′ core motif (Figure 4.4). Surface topology analysis of isolated TTF-1HD performed by limited proteolysis at neutral pH was in good agreement with the three-dimensional structure of the molecule as determined by NMR studies in acidic conditions [34–36]. Minor differences were detected in the C-terminal region of the protein, which, contrary to NMR data, showed no accessibility to proteases. Preferential cleavage sites in isolated HD gathered into a few specific regions of the protein, the most exposed segment being the N-terminal portion.

Following complex formation, a different cleavage site profile was obtained as a consequence of the marked protection effect exerted by the oligonucleotide on TTF-1HD. The recognition helix that had been found exposed in the isolated protein was totally protected following interaction, confirming that this region lies in the major groove of the DNA, establishing specific contacts with the nucleotides and with the sugar phosphate backbone. An increased accessibility of the C-terminal region was observed, indicating that

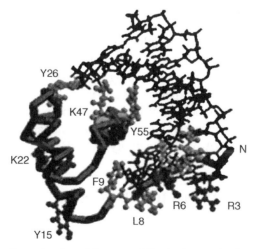

Figure 4.4   Three-dimensional model proposed for the interaction between TTF-1HD and its specific oligonucleotide.

the C-terminal portion of HD is displaced from the protein surface by the oligonucleotide molecule and adopts a disordered and flexible conformation. On the contrary, a large conformational change affected the N-terminal region of HD that was significantly shielded by the oligonucleotide as it intercalated into the minor groove of DNA. These findings suggest a direct effect of DNA binding on the molecular conformation of the entire TTF-1 protein; the displacement of the HD C-terminal tail might affect the C-terminal domain of the TTF-1 protein, causing this region to assume a biologically relevant conformation. The analysis of the TTF-1HD/DNA complex provided an experimental validation of the model proposed for the interaction on the basis of the HD structures described so far [37].

Very recently, a combination of limited proteolysis and extension cloning was used to identify the region of BRCA1 that binds specifically to four-way junction DNA, a property that potentially facilitates its role in the repair of DNA lesions by homologous recombination. Limited trypsinolysis of BRCA1 230–534 resulted in the production of a soluble 230–339 domain that was unable to bind DNA, suggesting that the binding activity, in part, resides within residues 340–534. A series of fragments extending from residues 340 were produced and tested for their ability to bind to four-way junction DNA in gel retardation assays, thus identifying residues 340–554 of BRCA1 as the minimal DNA binding region [38].

Besides studies on protein–DNA complexes, several reports have appeared on protein–RNA interactions probed by limited proteolysis and MS [39]. Human ribosomal protein L7a is a component of the major ribosomal subunit, contains two RNA-binding sites, and interacts in vitro with a G-rich RNA sequence. The topology of the L7a–RNA complex was investigated using a recombinant form of the protein encompassing residues 101–161 and a 30 mer poly(G) RNA oligonucleotide. Limited proteolysis experiments carried out on isolated L7a and the complex followed by ESMS analyses of the released fragments led to the identification of a protein region (KQRLLARAEK) that was shielded from proteases upon binding to poly(G). These data were confirmed by UV cross-linking experiments and provided experimental evidence that, in addition to the predicted RNA-binding domain (RNAB2), the domain previously shown to be essential for nucleolar accumulation of the human L7a r-protein also exerts RNA-binding activity (RNAB1).

4.6   PROBING PROTEIN–LIGAND INTERACTIONS

The presence or absence of a bound ligand can profoundly affect the susceptibility of a protein segment to limited proteolysis, although the mechanism by which these effects are achieved are not always clear. Indeed, the changes produced may be remote from the site of interaction of the protein with its substrate or cofactor. Hence, despite the little shielding effect exerted by the bound molecule, changes in the proteolytic profile upon ligand binding should be more likely ascribed to conformational variations induced by complex formation. Conformational studies of small molecules that bind to target proteins using limited proteolysis and mass spectrometry can then represent the fundamental experimental tool to obtain detailed structural information about binding sites and protein tertiary structure changes. This information is crucial to the development of therapeutic agents as well as of chemical sensors based on molecular recognition principles.

Shields and co-workers used time-resolved limited proteolysis experiments coupled to the high-throughput analysis capability of MALDI-TOF-MS to determine the binding site in a tetanus toxin C-fragment (51 kDa)–doxorubicin (543 Da) non covalent complex [40]. The relative ion abundances of peptides released from the limited proteolysis of tetanus toxin C-fragment (TetC) and the TetC–doxorubicin complex on a time course basis were carefully compared, revealing that the binding of doxorubicin induced a significant change in the surface topology of TetC. A decrease in ion abundance of specific peptides suggested doxorubicin obstructs the access of the protease to one or both termini of these peptides, identifying doxorubicin binding site(s). Conversely, increased abundance of peptide ions in the digest of the complex indicated an increase in accessibility to these sites, suggesting that doxorubicin not only binds to the surface but also induces a conformational change in TetC.

A combination of H/D exchange and limited proteolysis experiments coupled to mass spectrometry analysis was used to depict the conformation of HAMLET (human α-lactalbumin made lethal to tumor cells), a complex of α-lactalbumin with a specific fatty acid, C18:1, endowed with the peculiar biological function of inducing apoptosis in tumor and immature cells [41]. Conformational analyses of α-lactalbumin have been carried out by several independent groups and at least two different conformations of the protein have already been described under different physico chemical conditions, that is, at low pH (A state) and in the absence of Ca^{2+} (apo). Nevertheless, no structural details of HAMLET accounting for its molecular diversity from other forms of α-lactalbumin have been reported, as several techniques were not able to discriminate between HAMLET and apo-α-lactalbumin.

H/D exchange experiments clearly indicated that HAMLET and apo are indeed two different conformers of α-LA, as HAMLET incorporated a greater number of deuterium atoms compared to the apo and holo forms. Complementary proteolysis experiments showed that HAMLET and apo are both accessible to proteases in the β-domain with cleavages occurring at the same sites. However, substantial differences in the kinetics of enzymatic digestion at specific sites were observed. In particular, Asp37, Tyr50, and Phe53 were found to be by far more exposed in HAMLET than in apo. These results were supported by the analysis of the peptides generated from the peptic digest of deuterium-exchanged α-lactalbumin conformers, and suggested that binding of oleic acid might displace the central strand of the β-sheet, exposing Asp37, Tyr50, and Phe53.

4.7   PROBING AMYLOID FIBRIL CORE

Limited pro teolysis e xperiments i ntegrated wi th ma ss s pectrometric t ech-niques a re p articularly s uited f or s tudying pro tein co mplexes i n s olution. However, the ability of several proteases to exert their enzymatic activity in heterogeneous phase made it possible to investigate also ordered protein ag-gregate, such as amyloid fibrils.

Despite a l ack of significant se quence homology between the constituent proteins, amyloid fibrils appear to share common structures, as evidenced by very similar fibrillar morphologies and X-ray diffraction patterns indicative of cross-β-sheet structure [42, 43]. Although the gross morphology of amy-loid fibrils is fairly well understood, very little is known about how the con-stituent proteins thread into this apparent common folding motif. Since usual high-resolution methods of structure determination are not amenable to ad-dress the structures of these large, polydisperse, insoluble aggregates, struc-tural i nformation at l ower res olution ha s b ecome i ncreasingly de manding. Limited proteolysis studies are aimed at the definition of the protein region(s) involved in fibril formation, by discriminating the hydrophobic core of the fibrils from the still flexible protein segments protruding from the highly or-dered fibril structure.

In the 2001 Wetzel and co-workers reported studies on synthetic amyloid fibrils obtained in vitro from the Alzheimer's disease peptide Aβ (1–40) [44]. They performed a detailed kinetic analysis on the order of peptide releasing from the fibrils and found that the extreme N-terminal fragment of Aβ is gen-erated from fibrils as rapidly as it is from the Aβ soluble monomer, while other peptides are produced much more slowly. Furthermore, aggregated material isolated by centrifugation of intermediate digestion time points contained, in add ition to f ull-length mat erial, p eptides t hat pos sess mat ure C t ermini but truncated N termini. These data strongly suggested that the N-terminal region of Aβ is not involved in the β-sheet network of the amyloid fibril, while the C terminus is essentially completely engaged in the protective highly or-dered s tructure. H owever, w hile t he ma jority o f A β m olecules sho wed a n exposed N t erminus, abo ut 20 % o f t he p eptide e xhibited a n N t erminus shielded from proteolysis by the fibril structure. This heterogeneity could be explained by the involvement of the N-terminal segment in lateral association of protofilaments into the fibril structure, suggesting that the N-terminal region of Aβ, while not directly involved in the β-sheet network, may contribute to fibril stability by participating in protofilament packing.

β2-microglobulin (β2m) i s a s mall gl obular pro tein co nsisting o f e ight β-strands that aggregate to form classical amyloid fibrils in patients undergoing long-term hemodialysis. Amyloid fibrils essentially consist of wild-type β2-m and its truncated species ΔN6 β2-m lacking six residues at the amino terminus [45]. The limited proteolysis/mass spectrometry approach was employed to investigate the aggregation behavior of recombinant ΔN6 β2-m and to gain insights into the structure of β2-m fibrillar polymers [46]. Truncated ΔN6 β2-m showed a more flexible three-dimensional structure and a greater propensity to aggregate and to extend natural fibrils even at physiological pH [47].

Proteolytic analyses of fibrils obtained from both native β2-m and ΔN6 β2-m revealed that the central region of the protein was fully protected in the fibrillar form while the amino- and carboxy-terminal regions of β2-m became exposed to the solvent in the fibrils. These data indicate that amyloid fibrils of β2-m consist of an inaccessible core comprising residues 20–87 of the protein with the strands I and VIII not being constrained in the fibrillar polymer and exposed to the proteases. Moreover, proteolytic cleavages observed in vitro at Lys 6 and Lys 19 reproduce specific cleavages that have to occur *in vivo* to generate the truncated forms of β2-m occurring in natural fibrils.

Recently, Redford and co-workers demonstrated that β2-m amyloid fibrils obtained u nder d ifferent co nditions d isplay v ery d ifferent pro teolytic patterns. Amyloid fibrils were obtained from β2-m using different acidic conditions *in vitro* [48]. The morphology of these fibrils was tested by atomic force microscope investigations, resulting either in long and straight or in short or even in curved and nodular fibrils depending on the pH and ionic strength conditions used for fibril formation. Limited proteolysis experiments were individually carried out on fibrils with different morphology to investigate their conformational properties using pepsin as proteolytic probe. Because of t he w ide s pecificity o f p epsin, p eptide f ragments re leased f rom fi brils had to b e i dentified by tandem mass spectrometry following LC-MS/MS analyses.

According to their different morphology, fibrils originated distinct digestion patterns, demonstrating that the organization of the polypeptide chain in fibrils with different morphological features was considerably different, despite all the fibrils displaying a common cross-β architecture. While the curved, worm-like fibrils are relatively weakly protected from proteolysis, the long, straight fibrils showed only a single cleavage site in the N-terminal end, demonstrating that substantial refolding of the initially acid-denatured and unprotected state of β-2m occurred during assembly.

4.8 CONCLUSIONS

The examples presented illustrate the utility of combining proteolysis and mass spectrometry analyses in probing conformational studies of protein complexes and multicomponent proteinaceous assemblies. The experimental techniques that these approaches rely on are well established, efficient, readily accessible, and within reach of the entire biochemistry community. Moreover, as biomolecular mass spectrometry continues to expand as the technique of choice in bioanalytic studies, the necessary instrumentation is becoming widely available to most biochemical laboratories.

Structural investigations of protein molecular interactions face several different bottlenecks owing to the innate methodological requirements by X-ray diffraction and NMR studies. Although the limited proteolysis approach can only provide low-resolution data, the procedure described here is effective in probing the topology of interacting interfaces in a range of biological complexes. Moreover, limited proteolysis is amenable to the study of protein dynamics by monitoring conformational changes in protein structure under different experimental conditions and providing subtle structural details on highly flexible conformations such as transient species and partly folded intermediates. In this respect, it is particularly well suited to effectively supplement high-resolution techniques such as X-ray and/or NMR analyses in structural protein studies.

Modern views in structural protein studies have already recognized that the flexibility of proteins is instrumental not only to exploit catalytic activity but also to alter their biological functions, conferring peculiar properties to different conformations of the molecule. Therefore, protein studies that require information of the structural changes associated with different biological functions of the polypeptide chain will rely more and more on the kind of approaches described in this chapter.

REFERENCES

1. Wutrich, K. (1989). The development of nuclear magnetic resonance spectroscopy as a technique for protein structure determination. *Acc. Chem. Res.* **22**: 36–44.
2. Weber, P. C. (1991). Physical principle of protein crystallization. *Adv. Protein. Chem.* **41**: 1–36.
3. Fontana, A., Polverino de Laureto, P., De Fillipis, V., Scaramella, E., and Zambonin, M. (1997). Probing the partly folded states of proteins by limited proteolysis. *Fold Des.* **2**: 17–26.
4. Thornton, J. M., Edwards, M. S., Taylor, W. R., and Barlow, D. J. (1986). Location of "continuous" antigenic determinants in the protruding regions of proteins. *EMBO J.* **5**: 409–413.

5. Darby, N. J., Kemmink, J., and Creighton, T. E. (1996). Identifying and characterizing a structural domain of protein disulfide isomerase. *Biochemistry* **35**: 10517–10528.

6. Bantscheff, M., Weiss, V., and Glocker, M. O. (1999). Identification of linker regions and domain borders of the transcription activator protein NtrC from *Escherichia coli* by limited proteolysis, in-gel digestion, and mass spectrometry. *Biochemistry* **38**: 11012–11020.

7. Richards, F. M., and Vithayathil, P. J. (1959). The preparation of subtilisn-modified ribonuclease and the separation of the peptide and protein components. *J. Biol. Chem.* **234**: 1459–1464.

8. Jawhari, A., Boussert, S., Lamour, V., Atkinson, R. A., Kieffer, B., Poch, O., Potier, N., van Dorsselaer, A., Moras, D., and Poterszman, A. (2004). Domain architecture of the p62 subunit from the human transcription/repair factor TFIIH deduced by limited proteolysis and mass spectrometry analysis. *Biochemistry* **43**: 14420–14430.

9. Fontana, A., Fassina, G., Vita, C., Dalzoppo, D., Zamai, M., and Zambonin, M. (1986). Correlation between sites of limited proteolysis and segment mobility in thermolysin. *Biochemistry* **25**: 1847–1851.

10. Arnone, M. I., Birolo, L., Giamberini, M., Cubellis, M. V., Nitti, G., Marino, G., and Sannia, G. (1992). Limited proteolysis as a probe of conformational change in aspartate aminotransferase from *Sulfolobus solfataricus*. *Eur. J. Biochem.* **204**: 1183–1189.

11. Winslow, J. W., Van Amsterdam, J. R., and Neer, E. J. (1986). Conformations of the alpha 39, alpha 41, and beta-gamma components of brain guanine nucleotide-binding proteins. Analysis by limited proteolysis. *J. Biol. Chem.* **261**: 7571–7579.

12. Edwards, L. A., Tian, M. R., Huber, R. E., and Fowler, A. V. (1988). The use of limited proteolysis to probe interdomain and active site regions of beta-galactosidase (*Escherichia coli*). *J. Biol. Chem.* **263**: 1848–1854.

13. Corti, A., Sarubbi, E., Soffientini, A., Nolli, M. L., Zanni, A., Galimberti, M., Parenti, F., and Cassani, G. (1989). Epitope mapping of the anti-urokinase monoclonal antibody 5B4 by isolated domains of urokinase. *Thromb. Haemost.* **62**: 934–939.

14. Ramachandran, S., Richards-Sucheck, T. J., Skrzypczak-Jankun, E., Wheelock, M. J., and Funk, M. O. Jr. (1995). Catalysis sensitive conformational changes in soybean lipoxygenase revealed by limited proteolysis and monoclonal antibody experiments. *Biochemistry* **34**: 14868–14873.

15. Rao, L., Jones, D. P., Nguyen, L. H., McMahan, S. A., and Burgess, R. R. (1996). Epitope mapping using histidine-tagged protein fragments: application to *Escherichia coli* RNA polymerase sigma 70. *Anal. Biochem.* **241**: 173–179.

16. Morris, G. E. (1998). Epitope mapping. *Methods Mol. Biol.* **80**: 161–172.

17. Polverino de Laureto, P., Scaramella, E., Frigo, M., Wondrich, F. G., De Filippis, V., Zambonin, M., and Fontana, A. (1999). Limited proteolysis of bovine

alpha-lactalbumin: isolation and characterization of protein domains. *Protein Sci.* **8**: 2290–2303.

18. Brockerhoff, S. E., Edmonds, C. G., and Davis, T. N. (1992). Structural analysis of wild type and mutant yeast calmodulins by limited proteolysis and electrospray ionization mass spectrometry. *Protein Sci.* **1**: 504–516.

19. Cohen, S. L., Ferri-D Amari, A. R., Burley, K. S., and Chait, B. T. (1995). Probing the solution structure of the DNA-binding protein Max by a combination of proteolysis and mass spectrometry. *Protein Sci.* **4**: 1088–1099.

20. Zappacosta, F., Pessi, A., Bianchi, E., Venturini, S., Sollazzo, M., Tramontano, A., Marino, G., and Pucci, P. (1996). Probing the tertiary structure of proteins by limited proteolysis and mass spectrometry: the case of Minibody. *Protein Sci.* **5**: 802–813.

21. Atkinson, R. A., Joseph, C., Dal Piaz, F., Birolo, L., Stier, G., Pucci, P., and Pastore, A. (2000). Binding of alpha-actinin to titin: implications for Z-disk assembly. *Biochemistry* **39**: 5255–5264.

22. Kim, Y. J., Kim, Y. A., Park, N., Son, H. S., Kim, K. S., and Hahn, J. H. (2005). Structural characterization of the molten globule state of apomyoglobin by limited proteolysis and HPLC-mass spectrometry. *Biochemistry* **44**: 7490–7496.

23. Zhang, B., and Peng, Z. Y. (2002). Structural consequences of tumor-derived mutations in p16INK4a probed by limited proteolysis. *Biochemistry* **41**: 6293–6302.

24. Huang, S., Zou, X., Guo, P., Zhong, L., Peng, J., and Jing, G. (2005). Probing the subtle conformational state of N138ND2-Q106O hydrogen bonding deletion mutant (Asn138Asp) of staphylococcal nuclease using time of flight mass spectrometry with limited proteolysis. *Arch. Biochem. Biophys.* **434**: 86–92.

25. Shao, J., Irwin, A., Hartson, S. D., and Matts, R. L. (2003). Functional dissection of cdc37: characterization of domain structure and amino acid residues critical for protein kinase binding. *Biochemistry* **42**: 12577–12588.

26. Scaloni, A., Miraglia, N., Orrù, S., Amodeo, P., Motta, A., Marino, G., and Pucci, P. (1998). Topology of the calmodulin–melittin complex. *J. Mol. Biol.* **277**: 945–958.

27. Ikura, M., Spera, S., Barbato, G., Kay, L. E., Krinks, M., and Bax, A. (1991). Secondary structure and side-chain ^{1}H and ^{13}C resonance assignments of calmodulin in solution by heteronuclear multidimensional NMR spectroscopy. *Biochemistry* **30**: 9216–9228.

28. Barbato, G., Ikura, M., Kay, L. E., Pastor, R. W., and Bax, A. (1992). Backbone dynamics of calmodulin studied by ^{15}N relaxation using inverse detected two-dimensional NMR spectroscopy: the central helix is flexible. *Biochemistry* **31**: 5269–5278.

29. Chattopadhyaya, R., Meador, W. E., Means, A. R., and Quiocho, F. A. (1992). Calmodulin structure refined at 1.7 Å resolution. *J. Mol. Biol.* **228**: 1177–1192.

30. Casbarra, A., Dal Piaz, F., Ingallinella, P., Orru, S., Pucci, P., Pessi, A., and Bianchi, E. (2002). The effect of prime-site occupancy on the hepatitis C virus NS3 protease structure. *Protein Sci.* **11**: 2102–2112.

31. Failla, C., Tomei, L., and De Francesco, R. (1994). Both NS3 and NS4A are required for proteolytic processing of hepatitis C virus nonstructural proteins. *J. Virol.* **68**: 3753–3760.

32. Urbani, A., Biasiol, G., Brunetti, M., Volpari, C., Di Marco, S., Sollazzo, M., Orru, S., Dal Piaz, F., Casbarra, A., Pucci, P., Nardi, C., Gallinari, P., De Francesco, R., and Steinkühler, C. (1999). Multiple determinants influence complex formation of the hepatitis C virus NS3 protease domain with its NS4A cofactor peptide. *Biochemistry* **38**: 5206–5215.

33. Chauhan, H. J., Domingo, G. J., Jung, H. I., and Perham, R. N. (2000). Sites of limited proteolysis in the pyruvate decarboxylase component of the pyruvate dehydrogenase multienzyme complex of *Bacillus stearothermophilus* and their role in catalysis. *Eur. J. Biochem.* **267**: 7158–7169.

34. Viglino, P., Fogolari, F., Formisano, S., Bortolotti, N., Damante, G., Di Lauro, R., and Esposito, G. (1993). Structural study of rat thyroid transcription factor 1 homeodomain (TTF-1 HD) by nuclear magnetic resonance. *FEBS Lett.* **336**: 397–402.

35. Esposito, G., Fogolari, F., Damante, G., Formisano, S., Tell, G., Leonardi, A., Di Lauro, R., and Viglino, P. (1996). Analysis of the solution structure of the homeodomain of rat thyroid transcription factor 1 by ^{1}H-NMR spectroscopy and restrained molecular mechanics. *Eur. J. Biochem.* **241**: 101–113.

36. Scaloni, A., Monti, M., Acquaviva, R., Tell, G., Damante, G., Formisano, S., and Pucci, P. (1999). Topology of the thyroid transcription factor 1 homeodomain–DNA complex. *Biochemistry* **3**: 64–72.

37. Fogolari, F., Esposito, G., Viglino, P., Damante, G., and Pastore, A. (1993). Homology model building of the thyroid transcription factor 1 homeodomain. *Protein Eng.* **6**: 513–519.

38. Naseem, R., Sturdy, A., Finch, D., Jowitt, T., and Webb, M. (2006). Mapping and conformational characterization of the DNA-binding region of the breast cancer susceptibility protein BRCA1. *Biochem. J.* **395**: 529–535.

39. Russo, G., Cuccurese, M., Monti, G., Russo, A., Amoresano, A., Pucci, P., and Pietropaolo, C. (2005). Ribosomal protein L7a binds RNA through two distinct RNA-binding domains. *Biochem. J.* **385**: 289–299.

40. Shields, S. J., Oyeyemi, O., Lightstone, F. C., and Balhorn, R. (2003). Mass spectrometry and non-covalent protein–ligand complexes: confirmation of binding sites and changes in tertiary structure. *J. Am. Soc. Mass. Spectrom.* **14**: 460–470.

41. Casbarra, A., Birolo, L., Infusini, G., Dal Piaz, F., Svensson, M., Pucci, P., Svanborg, C., and Marino, G. (2004). Conformational analysis of HAMLET, the folding variant of human alpha-lactalbumin associated with apoptosis. *Protein Sci.* **13**: 1322–1330.

42. Makin, O. S., and Serpell, L. C. (2005). Structures for amyloid fibrils. *FEBS J.* **272**: 5950–5961.

43. Wetzel, R. (2002). Ideas of order for amyloid fibril structure. *Structure* **10**: 1031–1036.

44. Kheterpal, I., Williams, A., Murphy, C., Bledsoe, B., and Wetzel, R. (2 001). Structural features of the Abeta amyloid fibril elucidated by limited proteolysis. *Biochemistry* **40**: 11757–11767.

45. Stoppini, M., Arcidiaco, P., Mangione, P., Giorgetti, S., Brancaccio, D., and Bellotti, V. (2 000). D etection o f f ragments o f β2-microglobulin i n a myloid fibrils. *Kidney Int.* **57**: 349–350.

46. Monti, M., Principe, S., Giogetti, S., Mangione, P., Merlini, G., Clark, A., Bellotti, V., Amoresano, A., and Pucci, P. (2002). Topological investigation of amyloid fibrils obtained from β2-microglobulin. *Protein Sci.* **11**: 2362–2369

47. Esposito, G., Michelutti, R., Verdone, G., Viglino, P., Hernàndez, H., Robinson, C. V., Amoresano, A., Dal Piaz, F., Monti, M., Pucci, P., Mangione, P., Asti, L., Stoppini, M., Merlini, G., Ferri, G., and Bellotti, V. (2000). Removal of the N-terminal hexapeptide from human β2-microglobulin facilitates protein aggregation and fibril formation. *Protein Sci.* **9**: 831–845.

48. Myers, S. L., Thomson, N. H., Radford, S. E., and Ashcroft, A. E. (2006). Investigating the structural properties of amyloid-like fibrils formed in vitro from beta(2)-microglobulin us ing l imited prot eolysis a nd e lectrospray i onisation mass spectrometry. *Rapid Commun. Mass Spectrom.* **20**: 1628–1636.

5

CHEMICAL CROSS-LINKING AND MASS SPECTROMETRY FOR INVESTIGATION OF PROTEIN–PROTEIN INTERACTIONS

ANDREA SINZ

Institute of Pharmacy, Martin Luther University Halle–Wittenberg, Germany

5.1 I ntroduction

5.2 C ross-Linking Strategies

    5.2.1 B ottom–Up Approach

    5.2.2 T op–Down Approach

5.3   Functional Groups of Cross-Linking Reagents: Reactivities

    5.3.1 A mine-Reactive Cross-Linkers

    5.3.2 S ulfhydryl-Reactive Cross-Linkers

    5.3.3. P hotoreactive Cross-Linkers

5.4 C ross-Linker Design

    5.4.1 H omobifunctional Cross-Linkers

    5.4.2 Het erobifunctional Cross-Linkers

    5.4.3 Z ero-Length Cross-Linkers

    5.4.4 T rifunctional Cross-Linkers

5.5   Mass Spectrometric Analysis of Cross-Linked Products

    5.5.1   Bottom–Up Analysis by MALDI-MS

    5.5.2   Bottom–Up Analysis by ESI-MS (LC/MS)

    5.5.3   Bottom–Up and Top–Down Analysis by ESI-FTICR-MS

5.6 I dentification of Cross-Linked Products

5.7   Computer Software for Data Analysis

Mass Spectrometry of Protein Interactions Edited by Kevin M. Downard
Copyright © 2007 John Wiley & Sons, Inc.

5.8   Conclusions and Perspectives
Abbreviations
Acknowledgments
References

5.1   INTRODUCTION

Closely related to studying the function of a protein is the identification and characterization of its interaction partners. In those cases where high-resolution methods are applicable for structural analysis, such as X-ray crystallography and nuclear magnetic resonance (NMR) spectroscopy, the solved three-dimensional structure of a protein gives insights into stable interactions within a protein complex. Another strategy to obtain structural information of a protein assembly consists in the incorporation of chemical reagents (cross-linkers) into the complex. The location of the created cross-links imposes a distance constraint on the location of the respective side chains and allows drawing conclusions on the distances of a protein complex structure [1–3].

Analysis of cross-linked peptides by mass spectrometry (MS) makes use of several advantages associated with MS analysis. The mass of the protein or the protein complex under investigation is theoretically unlimited because it is the proteolytic peptides that are analyzed, where a bottom–up strategy is employed. Analysis is generally fast and, in favorable circumstances, requires only femtomole amounts of total protein. Furthermore, it is possible to gain insights into three-dimensional structures of proteins in solution and flexible regions are readily identified. One of the greatest advantages of chemical cross-linking is that membrane proteins, which are otherwise difficult to study using established methodology, are amenable to analysis. The broad range of specificities available for cross-linking reagents toward certain functional groups, such as primary amines, sulfhydryls, or carboxylic acids, and the wide range of distances that different cross-linking reagents can bridge offer the possibility to perform a wide variety of experiments [4].

However, despite the straightforwardness of the cross-linking approach, the identification of the cross-linked products can be hampered by the complexity of the reaction mixtures. Several strategies have been employed to enrich cross-linker-containing species by affinity chromatography or to facilitate the identification of the cross-linked products, for example, by using isotope-labeled cross-linkers or proteins, fluorogenic or fluorescent cross-linkers, or cleavable cross-linkers.

Here, the most popular cross-linking reagents for protein structure analysis are described and an overview is given of the currently available strategies that employ chemical cross-linking and different mass spectrometric techniques.

5.2   CROSS-LINKING STRATEGIES

Two general strategies exist for chemical cross-linking in combination with a mass spectrometric analysis of the created products: the bottom–up and top–down approaches [5]. In the following, both strategies are described and compared to each other with respect to their specific strengths and limitations.

5.2.1   Bottom–Up Approach

In the bottom–up approach, the protein reaction mixture is enzymatically digested after the cross-linking reaction, and mass spectrometric identification of the cross-linked products is performed—based on the resulting proteolytic peptides (Figure 5.1a). The bottom–up approach has been applied to map protein interfaces, but it has also proved especially valuable to determine low-resolution three-dimensional structures of proteins [1, 3]. The most important prerequisite to successfully conduct cross-linking experiments is a detailed description of the respective amino acid sequences of the proteins under investigation. Full sequence coverage should be envisioned to fully characterize the protein with respect to possible amino acid variants, post-translational modifications, or splice variants. When conducting cross-linking reactions, control samples must be included, to which no cross-linker is added, in order to exclude the formation of any nonspecific aggregates. Moreover, cross-linker concentrations, reaction times, and buffer pH must be optimized to achieve a high yield of cross-linked product, while not disrupting the three-dimensional protein structures by introducing too many cross-links per molecule.

After the cross-linking reaction, one-dimensional gel electrophoresis (SDS-PAGE) and MALDI-TOF (matrix-assisted laser desorption/ionization time-of-flight) MS analysis of the reaction mixture can be used to check for the extent of cross-linked product formation and to optimize the reaction conditions. After the cross-linking reaction, there are several ways to isolate the cross-linked proteins from the reaction mixture. If SDS-PAGE of the cross-linking reaction mixture is performed, the band of the cross-linked protein or the cross-linked protein complex is excised from the one-dimensional gel and subjected to enzymatic *in-gel* digestion (Figure 5.1a). Based on the staining intensity of the respective gel bands, the amount of cross-linked product formation is approximated. Alternatively, the cross-linked protein or protein complex is separated from the reaction mixture by

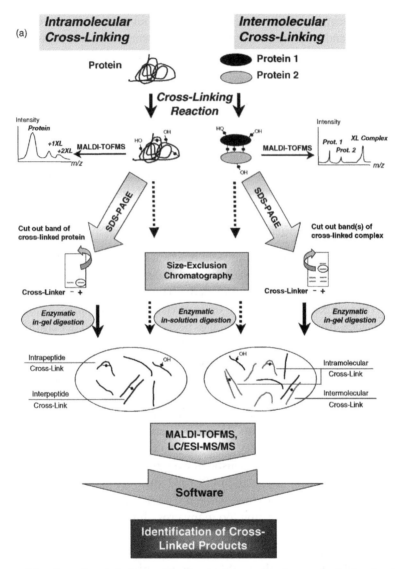

Figure 5.1   General analytical strate gies for protein structure characterization by chemical cross-linking and FTICR mass spectrometry (a) Bottom–up approach and (b) top–down approach. Figure adapted from Sinz [5], with permission of Springer Science and Business Media.

size-exclusion chromatography, and the digestion is performed in the solution (Figure 5.1a). We found that an *in-solution* digestion of the cross-linked proteins is far more efficient than *in-gel* digestion, where ~80% of the protein is lost. The resulting highly complex peptide mixtures generated from enzymatic digestion contained unmodified peptides of the protein(s), peptides modified by partially hydrolyzed cross-linker, intramolecular (inter- and intrapeptide) cross-linked products between peptides originating from

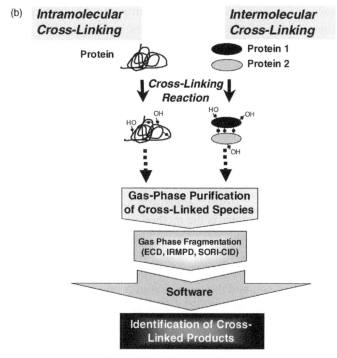

Figure 5.1   *(Continued)*

one protein, as well as intermolecular cross-linked products between peptides from different proteins (Figure 5.1a). Peptide mixtures that originated
from proteolytic digestion of cross-linking reaction mixtures were analyzed
by MALDI or ESI mass spectrometry. The cross-linked peptides were assigned in the mass spectra, using customized software programs, such as
the GPMAW software (available at h ttp://welcome.to/gpmaw) [6]. Based
on signals in the mass spectra of cross-linking mixtures, but not in those of
control samples from non-cross-linked proteins, cross-linked products were
identified to ultimately provide further information on the spatial distances
between functional groups of the proteins under investigation.

One o f t he i nherent pro blems o f t he bo ttom–up s trategy i s t hat l arge
peptides a re re gularly creat ed f rom cross-linked pro teins d uring e nzymatic
proteolysis d ue to a h igh f requency of m issed c leavages. M issed c leavages
occur b ecause t he m ost co mmonly e mployed cros s-linking reag ents react
with primary amine groups at lysine residues and the N termini of proteins,
and trypsin—the most commonly used proteolytic enzyme—will not cleave
C terminal to a modified lysine residue. Another limitation of the bottom–up
approach is that cross-linked products with low charge states are frequently
created during electrospray ionization due to a loss of positive charge after
 modification of the ε-amino groups of lysine residues; that modification might
cause l arge p eptides n ot to b e de tected. M oreover, t he n umber o f p eptides

with the same nominal mass, but different amino acid sequence, increases with the rising number of amino acid residues in the peptide. Thus, mass spectrometric techniques, which yield high-resolution and high mass accuracy data, and moreover allow MS/MS fragmentation of large peptides, are essential for an unambiguous assignment of cross-linked products.

5.2.2 Top–Down Approach

The top–down approach presents the most direct technique to analyze cross-linked products. Here, the cross-linked proteins are analyzed intact rather than being digested before the mass spectrometric analysis [7] (Figure 5.1b). Electrospray ionization Fourier-transform ion-cyclotron resonance (ESI-FTICR) mass spectrometry is the method of choice for this kind of analysis. However, the top–down approach has so far been exclusively employed to determine low-resolution three-dimensional structures of relatively small proteins [8–10]. The cross-linking reaction mixture is presented to the FTICR mass spectrometer, and the cross-linked product is isolated in the ICR cell before it is fragmented using SORI-CID (sustained off-resonance irradiation collision-induced dissociation), IRMPD (infrared multiphoton dissociation), or ECD (electron capture dissociation) (Figure 5.1b). Instruments, such as the novel commercially available hybrid FTICR mass spectrometers, additionally offer the possibility to select and fragment ions prior to the ICR cell. Determination of the accurate mass of the intact cross-linked product provides hints on the number of incorporated cross-linker molecules as well as on the number of modifications caused by partially hydrolyzed cross-linkers. The top–down approach presents some advantages over the bottom–up approach in that it eliminates the need to separate the reacted protein from the cross-linking reaction mixture before the mass spectrometric analysis, as this separation is accomplished by a "gas-phase purification" in the mass spectrometer. After fragmentation of the cross-linked protein in the FTICR mass spectrometer, assignment of the cross-linked products is performed manually or by customized software programs (MS2PRO, available at http://roswell.ca.sandia.gov/~mmyoung/). Electron capture dissociation (ECD) seems to be especially favorable in conjunction with FTICR-MS because it allows a comprehensive fragmentation of large peptides while post-translational modifications are kept intact. One limitation of the top–down approach is that analyses of large protein assemblies are difficult to perform. In the case of characterizing bovine rhodopsin [10], the protein was proteolyzed into large peptide fragments, using cyanogen bromide, which cleaves at the C-terminal site of methionine residues, before ESI-FTICR-MS/MS experiments were conducted in a top–down fashion. That combination of bottom–up and top–down analysis will most likely become the strategy with the greatest potential for a rapid and efficient analysis of a wide variety of cross-linking reaction mixtures.

5.3   FUNCTIONAL GROUPS OF CROSS-LINKING REAGENTS: REACTIVITIES

The variety of cross-linking reagents has increased dramatically during the past 25 years, and today, a wide variety of reagents are commercially available that possess different spacer lengths and reactivities (product catalogs available, e.g., at http://www.piercenet.com/ and http://www.trc-canada.com/). In the following, the most widely used classes of cross-linking reagents are described with respect to their specific strengths and limitations.

5.3.1   Amine-Reactive Cross-Linkers

N-hydroxysuccinimide (NHS) esters are probably the most widely applied principle to create reactive acylating reagents (Scheme 5.1a). Thirty years

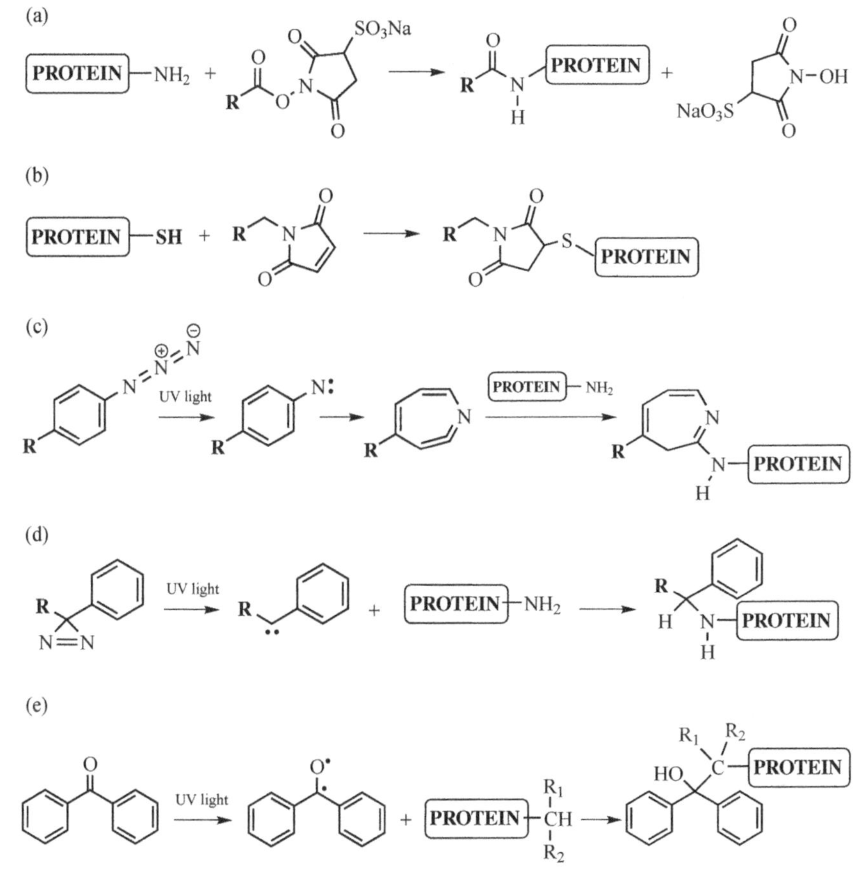

Scheme 5.1   Reaction schemes of the most commonly used reagents for cross-linking of proteins. (a) NHS esters (amine-reactive), (b) maleimides (sulfhydryl-reactive), (c) aryl azides (photoreative), (d) diazirines (photoreactive), (e) benzophenones (photoreactive) (f) "zero-length" cross-linker EDC in combination with sulfo-NHS (amine/carboxylic acid-reactive).

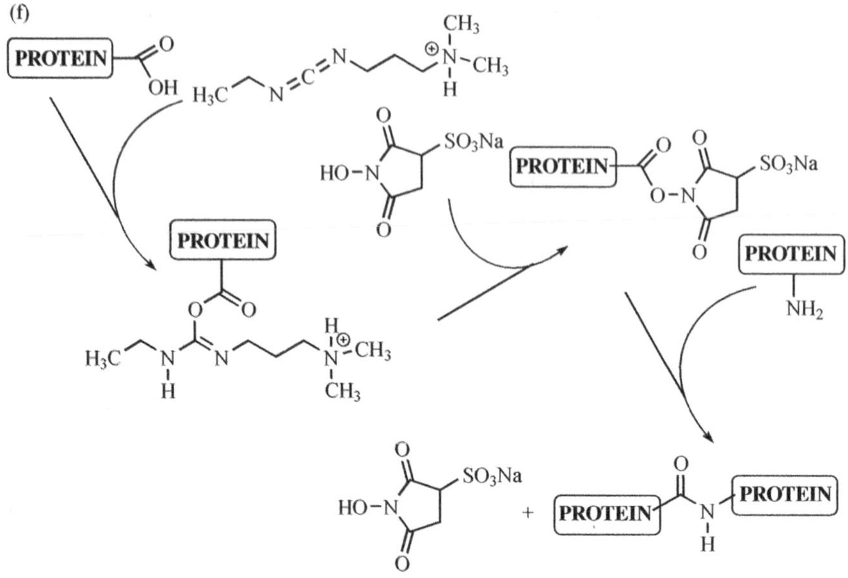

(f)

Scheme 5.1 *(Continued)*

ago, NHS esters were introduced as homobifunctional, highly amine-reactive, cross-linking reagents [11, 12]. NHS esters react with nucleophiles to release the NHS or sulfo-NHS group and to create stable amide and imide bonds with primary or secondary amines, such as free the N terminus and ε-amino groups in lysine side chains of proteins (Scheme 1a). NHS esters exhibit half-lives on the order of hours under physiological pH conditions (pH 7.0–7.5) with hydrolysis and amine reactivity increasing when the pH is raised [4, 12–14].

When conducting cross-linking experiments with NHS esters in our laboratory, we frequently observe cross-linked products of serine hydroxyl groups in MS analysis, so the created esters are apparently sufficiently stable in aqueous solution. This observation is confirmed by a carefully conducted study, using the homobifunctional, amine-reactive, and cleavable cross-linker DTSSP (3,3'-dithiobis[sulfosuccinimidyl propionate]) [15]. Reaction products of several model peptides with DTSSP were analyzed by ESI-QqTOF mass spectrometry and reaction products were confirmed by tandem MS experiments. The NHS ester DTSSP was found to react unexpectedly with contaminant ammonium ions in the buffer solution and with serine and tyrosine residues in addition to the desired reactions with lysine residues and the N terminus [15]. Another study describes the formation of stable products when NHS esters reacted with primary amines and tyrosine OH groups [16]. Under acidic conditions (pH 6.0), the NHS esters were found to react preferentially with the N terminus and tyrosine hydroxyl groups; however, under alkaline conditions (pH 8.4), they were found to react preferentially with the N terminus and lysine amine groups. These findings underline the urgent need to conduct further research on the reactivities of NHS esters, and cross-linking reagents in general; those studies have unfortunately been largely neglected so far.

5.3.2   Sulfhydryl-Reactive Cross-Linkers

The problem when targeting SH groups of cysteines is their possible involvement in disulfide-bond formation. Reduction of disulfide bonds in order to create free SH groups implies the danger of distorting the three-dimensional protein structure of the respective protein. Maleic acid imides—or maleimides—are a widely used reactive group in many heterobifunctional cross-linking reagents (Scheme 5.1(b)). Maleimide reactions are sulfhydryl specific in the pH range between 6.5 and 7.5 [47–20], and at pH 7 the reaction with maleimides proceeds approximately 1000 times faster with sulfhydryls than with amines.

5.3.3   Photoreactive Cross-Linkers

Photoreactive cross-linkers are induced to react with target molecules by exposure to ultraviolet (UV) light. The ideal photoreactive agent should be of high reactivity, capable of indiscriminately inserting into any type of residue, stable in the dark, and highly susceptible to light of a wavelength that does not cause any photolytic damage to the biological sample. Moreover, the reaction with proteins should lead to stable and unique products in order to enable their isolation, purification, and subsequent mass spectrometric analysis. By far, the largest number of photoreactive reagents is based on nitrene or carbene chemistry with the photolabile precursors being azides, diazirines, diazo compounds, and benzophenones.

Phenyl- and nitro-substituted phenyl azides have played a leading role in many areas of photochemical labeling, and currently they constitute the most commonly employed class of photoreactive cross-linkers (Scheme 5.1(c)). Upon photolysis, phenyl azide groups form short-lived nitrenes that can insert nonspecifically into chemical bonds of target molecules, including addition reactions at double bonds and insertion reactions into active hydrogen bonds at C—H and N—H sites [21]. The major disadvantages of aryl azides, however, are that they are activated by short-wavelength UV irradiation (<280 nm) and that the intermediately created nitrene may react nonuniformly.

Diazirines are remarkably stable to a variety of chemical conditions and are efficiently photolyzed at wavelengths of approximately 360 nm to generate a highly reactive carbene that can insert into a heteroatom—H or C—H bond (Scheme 5.1(d)). Unfortunately, photolysis of diazirines might lead to diazo isomers, which present strongly alkylating species that are responsible for undesired reactions in the dark [22].

A completely different photochemistry compared to aryl azides or diazirines is exhibited by benzophenones [23–25], which create a biradical upon irradiation. Subsequently, the oxygen radical abstracts a hydrogen radical from a bond of the reaction partner (Scheme 5.1(e)). The alkyl radicals created react by forming a new C—C bond between the photophor and the receptor protein. In contrast to diazirine compounds, activation of benzophenones does not proceed according to a photodissociative mechanism and is therefore reversible.

5.4   CROSS-LINKER DESIGN

5.4.1   Homobifunctional Cross-Linkers

Homobifunctional cross-linking reagents contain identical functional groups at both reactive sites, which are connected with a carbon-chain spacer that bridges a defined distance and thus allows identical functional groups of proteins to be cross-linked.

The main disadvantage of homobifunctional reagents is their susceptibility to create a wide range of poorly defined products [26]. The cross-linking reagent reacts initially with a protein molecule to form an intermediate, which could react with a second protein molecule to create a high-molecular-weight aggregate or which, alternatively, could react intramolecularly with a neighboring functional group on the same polypeptide chain. In order to check for high-molecular-weight aggregates due to intermolecular cross-linking, one-dimensional gel electrophoresis as well as a rapid mass spectrometric screening (e.g., by MALDI-TOF-MS) should be performed to establish the optimum cross-linking reaction conditions for the different cross-linking reagents. Special caution has to be applied not to disturb the three-dimensional structure of the proteins by excessive cross-linking; but, on the other hand, sufficient amounts of cross-linked products have to be created to allow for a subsequent mass spectrometric detection.

Single-step reaction procedures, using homobifunctional cross-linking reagents, in which all reagents are added at the same time to the reaction mixture, pose the greatest potential to form a multitude of different cross-linked products. To overcome that limitation, two-step protocols have been developed, in which one of the proteins is first reacted with the cross-linker in order to form an "activated" protein. After the reaction, excess reagent is removed and the activated protein is mixed with the second protein for cross-linking [4].

5.4.2   Heterobifunctional Cross-Linkers

Heterobifunctional cross-linking reagents contain two different reactive groups that target different functional groups on proteins. Those cross-linking reagents are used to cross-link proteins favorably in two- or three-step protocols in order to minimize the degree of high-molecular-weight aggregate formation. For example, an NHS ester/photoreactive heterobifunctional cross-linker can be applied for reaction with a protein amine group at its NHS ester function, whereas the photoreactive group is stable until it is exposed to high-intensity UV light. After a purification step, the photoreactive function of the cross-linker is brought to reaction with a protein CH or NH group by UV irradiation.

5.4.3   Zero-Length Cross-Linkers

Carbodiimides a re s o-called "zero-length" cros s-linking reag ents b ecause they do not introduce a spacer chain into the protein. They are used to mediate amide bond formation between spatially close ($<$ ~3 Å) groups; for example, a carboxylate and an amine group or a phosphate and an amine group [27–29]. Carbodiimide-mediated amide formation occurs effectively between pH 4.5 and 7.5. EDC (1-ethyl-3-(3-dimethylaminopropyl)carbodiimide) is the most popular representative [4]. EDC is mostly applied in combination with sulfo-NHS (*N*-hydroxysulfosuccinimide) [30]. The purpose of add ing sulfo-NHS to EDC is to increase the stability of the active intermediate, which ultimately reacts with the amine group according to S cheme 5.1(f). ED C/sulfo-NHS-coupled reactions are highly efficient and usually increase the yield of cross-linked product formation compared to EDC alone [30].

5.4.4   Trifunctional Cross-Linkers

The trifunctional cross-linker approach combines elements of the heterobifunctional cross-linker concept with the incorporation of an additional third functional group that can be specifically linked to a t hird protein or is used for a ffinity purification of c ross-linker-containing s pecies, in c ase a b iotin moiety is incorporated [31, 32]. In an excellent study conducted by Trester-Zedlitz et al. [31] five trifunctional cross-linking reagents were synthesized, including two or more of the following groups: an amine-reactive NHS ester, a sulfhydryl-reactive ma leimide, a ph otochemically reactive benzophenone, an i sotope t ag, a b iotin ha ndle, a nd/or a b ase-labile es ter c leavage s ite.

The incorporation of isotope labels into a biotinylated, trifunctional cross-linker is one of the most promising strategies for designing novel cross-linking reagents.

5.5   MASS SPECTROMETRIC ANALYSIS OF CROSS-LINKED PRODUCTS

The main challenges to identify cross-linked products by mass spectrometry arise from the high complexity of the reaction mixtures. The soft-ionization techniques MALDI (matrix-assisted laser desorption/ionization) [33] and ESI (electrospray ionization) [34] are the predominantly employed methods to analyze cross-linking mixtures of proteins.

5.5.1   Bottom–Up Analysis by MALDI-MS

The mechanisms of ion formation in MALDI are a subject of continuing research [35–38]. MALDI generated a great demand for a mass analyzer ideally suited to be used in conjunction with a pulsed ion source, such as the time-of-flight (TOF) analyzer. The performance of TOF instruments has increased tremendously during the past few years. Two true tandem TOF instruments have become commercially available [39, 40] and are likely to be beneficial to analyze cross-linked products. By conducting MS/MS experiments of cross-linked peptides, sequence information of the cross-linked peptides and information on the sites of cross-linking both become available [41]. MALDI-TOF-MS has been applied in numerous studies to analyze cross-linking reaction mixtures; for example, as described by References [1, 24, 31, 42–56]. In one report, MALDI-quadrupole ion trap (QIT) mass spectrometry has been employed to identify cross-linked products [57].

5.5.2   Bottom–Up Analysis by ESI-MS (LC/MS)

In ESI, liquids are sprayed in the presence of a strong electric field, forming small, highly charged droplets. ESI requires a sample that is devoid of nonvolatile salts and detergents to obtain the highest sensitivity as well as careful optimization of electrospray conditions for the specific compound under investigation. Miniaturization of the electrospray technique (nano-electrospray), by applying narrower spray capillaries, results in smaller droplets, reduced flow rates, and improved sensitivity [58, 59]. The peptide mixture is usually introduced into the mass spectrometer by a separation technique such as liquid chromatography (LC) or capillary electrophoresis (CE). Complex peptide mixtures are mostly separated by reversed-phase high-performance liquid chromatography (RP-HPLC). ESI-MS/MS is the method of choice to obtain further information on the amino acid sequence of cross-linked peptides as well as on the cross-linked amino acids. MS experiments using ESI-QIT, ESI-QqTOF, or ESI-TOF instruments are frequently employed for a detailed analysis of cross-linked products [15, 24, 31, 46–50, 56, 60–66].

5.5.3   Bottom–Up and Top–Down Analysis by ESI-FTICR-MS

For a confident assignment of cross-linker-containing species, the application of high-resolution and high-mass-accuracy methods, such as FTICR mass spectrometry, is a valuable prerequisite [67, 68]. With its ultra-high resolution and mass accuracy FTICR-MS offers the possibility of unambiguously identifying cross-linked species solely based on accurate mass measurements [69, 70].

In FTICR mass spectrometers, precursor-ion selection is accomplished by storing the ions of interest, whereas all others are ejected by means of a suitably tailored excitation pulse; for example, using the SWIFT technique [71]. MS/MS experiments are performed, using SORI-CID (sustained off-resonance irradiation collision-induced dissociation) [72], IRMPD (infrared multiphoton dissociation) [73], or ECD (electron capture dissociation) [74, 75]. Alternatively, with FTICR mass spectrometers that possess a quadrupole or a linear ion trap in front of the ICR cell, it is possible to conduct MS/MS experiments prior to the ICR cell. ESI-FTICR mass spectrometry has been coupled *on-line* with capillary and nano-liquid chromatography for high-throughput peptide identification with high sensitivity [76, 77]. In our group, we have been using nano-HPLC/nano-ESI-FTICR-MS to analyze cross-linking mixtures created from intramolecular cross-linking of proteins [78] and from intermolecular cross-linking of protein–peptide complexes, using a bottom–up approach [69, 79–81]. Investigation of the calcium-dependent complex between calmodulin and a 26 amino acid peptide derived from the skeletal muscle myosin light-chain kinase (M13) yielded several distance constraints that were in agreement with the published NMR structure of the calmodulin–M13 complex (Figure 5.2). Cross-linking experiments yielded cross-linked products between lysine residues at positions 18 and 19 of M13 and the central α-helix of calmodulin. The deconvoluted ESI-FTICR mass spectrum of the tryptic peptide mixture from the calmodulin–M13 (1:1) complex, cross-linked with BS³, is presented in Figure 5.3.

ESI-FTICR mass spectrometry is the method of choice for top–down analyses. However, it has so far only been successfully employed to assign cross-linked products from intramolecular cross-linking of small proteins, such as rhodopsin [10] or ubiquitin [8].

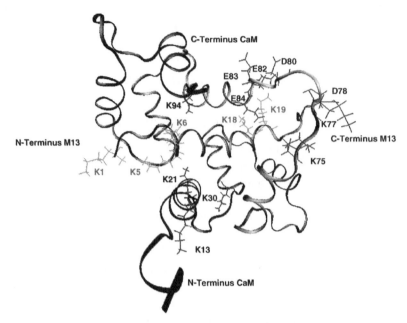

Figure 5.2   NMR structure of the calmodulin–M13 complex according to pdb entry "2BBM." The side chains of lysines and acidic amino acids that are potentially involved in cross-linking are indicated. CaM: calmodulin. Reprinted with permission from Kalkhof et al. [79].

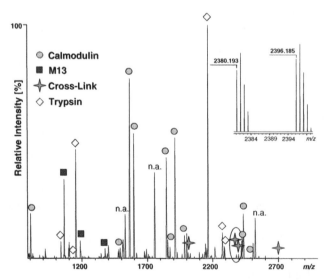

Figure 5.3   Deconvoluted ESI-FTICR mass spectrum of the tryptic peptide mixture from the calmodulin–M13 (1:1) complex cross-linked with BS³ (100-fold molar excess over protein/peptide concentration) at a reaction time of 60 min. The inset shows the magnified signal of a cross-linked product between calmodulin residues 75–86 and M13 residues 19–26 at m/z 2380.193, which was also detected with one methionine residue oxidized at m/z 2396.185; n.a.: signal not assigned. Reprinted with permission from Kalkhof et al. [79].

5.6 IDENTIFICATION OF CROSS-LINKED PRODUCTS

As mentioned earlier, mass spectrometric identification of cross-linked products can be hampered by the inherent complexity of the cross-linking reaction mixtures. The employed strategies to search for cross-linked products in complex mixture—comparable to the search for the "needle in the haystack"—utilize:

- Isotope-labeled cross-linkers or proteins
- Cross-linkers with affinity tags
- F luorogenic cross-linkers
- Chemically or MS/MS cleavable cross-linkers

The strategy using isotope-labeled cross-linkers will be described in the following in more detail. The application of 1:1 (w/w) mixtures of stable isotope-labeled cross-linking reagents allows low-abundance cross-linked peptides to be easily detected by their distinctive isotopic patterns after enzymatic digestion [46, 47]. We employed the homobifunctional, amine-reactive NHS esters BS^3 (*bis*[sulfosuccinimidyl]suberate) a nd B S^2G (*bis*[sulfosuccinimidyl]glutarate) (Table 5.1) to map the calcium-dependent complex between calmodulin and a peptide derived from the C terminus of adenylyl cyclase 8 [80]. BS^3 and BS^2G were employed to conduct cross-linking reactions as 1:1 mixtures of nondeuterated and four-times deuterated derivatives (d_0/d_4) (Table 5.1). Thus, an additional criterion for the identification of cross-linked products was introduced, because e very s pecies t hat co ntained o ne cros s-linker m olecule e xhibited a

TABLE 5.1   The Homobifunctional, Amine-Reactive, Isotope-Labeled Cross-linkers BS^2G, DSA, and BS^3 with Their Respective Spacer Chain Lengths

Cross-linker	Chemical Structure	Spacer Length
BS^2G-d_0/d_4 *Bis*(sulfosuccinimidyl) glutarate-d_0 *Bis*(sulfosuccinimidyl)2,2,4,4 glutarate-d_4		7.7 Å
DSA-d_0/d_8 Disuccinimidyl adip ate-d_0 Disuccinimidyl adip ate-d_8		8.9 Å
BS^3-d_0/d_4 *Bis*(sulfosuccinimidyl) suberate-d_0 *Bis*(sulfosuccinimidyl) 2,2,7,7 suberate-d_4		11.4 Å

doublet with a mass difference of 4.025 amu in the deconvoluted ESI-FTICR mass spectra. The isotope-labeled reagents BS^3, BS^{2}G, and DSA (disuccinimidyl adipate) (Table 5.1) were employed for mapping the interaction region in the 100 kDa tetrameric complex between annexin A2 and the S100A10 protein. In Figure 5.4, the ESI-FTICR mass spectrum of a tryptic peptide mixture from a cross-linking reaction mixture is shown, in which the monomeric annexin A2 protein has been intramolecularly cross-linked using the homobifunctional, amine-reactive, cross-linker DSA (Table 5.1). The cross-linker was employed as 1:1 mixture of nondeuterated and eight-times deuterated derivative in order to facilitate identification of cross-linker-containing species, which exhibit low signal intensities in the mass spectra (Figure 5.4).

In a recently published study, the homobifunctional NHS ester EGS (ethylene glycol bis(succinimidylsuccinate)) was employed in a 1:1 mixture of nondeuterated and 12-times deuterated derivative [41]. EGS possesses the additional advantage to be cleaved by ammonia, causing the release of cross-linker-containing peptides and thus allowing a rapid identification of cross-linked amino acids based on MS/MS experiments.

Another promising strategy using ^{18}O labeling of the homobifunctional amine-reactive cross-linker BS^3 has been reported [82]. A bis-^{18}O-labeled cross-linker was synthesized, so that proteins might be cross-linked by a defined mixture of unlabeled and ^{18}O-labeled reagent to produce two sets

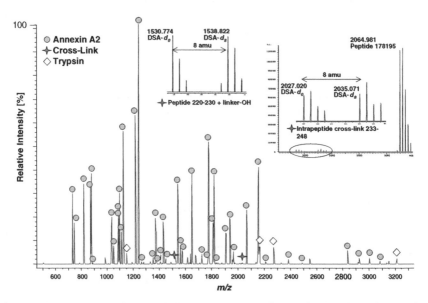

Figure 5.4 Deconvoluted ESI-FTICR mass spectrum of a tryptic digestion mixture of an annexin A2 monomer gel band, intramolecularly cross-linked with DSA-d_0/d_8 (100-fold excess over protein concentration, 30 min reaction time). Signals of cross-linker-containing species are shown enlarged.

of signals for cross-linked products that are separated by 4 amu. Peptides that are modified by partially hydrolyzed cross-linker are easily resolved by executing a simultaneous experiment that used unlabeled BS^3 in the presence of $H_2{}^{18}O$, resulting in a 2 amu mass shift for singly modified peptides. Reactions were monitored by MALDI-TOF-MS and ESI-TOF-MS analysis. Compared to deuterated cross-linkers, that strategy possessed the advantage that ^{18}O-labeled cross-linkers do not show any isotope effects in LC/MS analysis, whereas deuterated cross-linkers might exhibit slightly different retention times compared to their nondeuterated counterparts.

5.7   COMPUTER SOFTWARE FOR DATA ANALYSIS

The greatest deficits of employing chemical cross-linking and MS analysis consist presumably in the lack of computer software that can effectively analyze the enormous complexity of the reaction mixtures. All of the currently existing programs exhibit their specific limitations; thus, most of the cross-linked products must be manually assigned in the mass spectra. Intrapeptide cross-links or peptides that have been modified by a partially hydrolyzed cross-linker are easily identified by performing standard "in silico digestion" procedures; for example, by using the ExPASy Proteomics Tool "FindPept" (http://www.expasy.ch). Programs that are currently available as Web server versions for free access are the Automated Spectrum Assignment Program (ASAP), the MS2Assign, the MS2PRO software (available at http://roswell. ca.sandia.gov/~mmyoung/) [1, 8, 82, 83], and the SearchXLinks Version 3.3.3 software (available at th ttp://www.searchxlinks.de/cgi-bin/home.pl) [39, 84]. A novel software, termed VirtualMSLab, was presented recently, which is freely available upon request [85]. A commercially available program that allows calculating cross-linked products from one or two proteins is the GPMAW program (General Protein/Mass Analysis for Windows, Version 7.0) (http://welcome.to/gpmaw) [6].

5.8   CONCLUSIONS AND PERSPECTIVES

Structural analysis of proteins by chemical cross-linking combined with a mass spectrometric analysis of the created cross-linked products is a rapidly developing area. Recent technological advances in the field of mass spectrometry are likely to benefit the analysis of the complex mixtures created by chemical cross-linking. For confidently assigning cross-linker-containing species, the application of high-resolution and high-mass-accuracy methods, which additionally allow obtaining MS/MS data, is essential.

One has to be aware, however, that chemical cross-linking combined with mass spectrometry is still in its infancy to become a generally applicable technique to rapidly characterize protein interfaces. Clearly, improvements are needed in synthesizing novel cross-linking reagents, in better understanding the reactivities of different cross-linkers, in developing innovative strategies for facilitated MS detection, and in improving computer software for automated data analysis.

ABBREVIATIONS

BS^{2}G	*Bis*(sulfosuccinimidyl)glutarate
BS^3	*Bis*(sulfosuccinimidyl)suberate
CE C	apillary electrophoresis
CID Co	llision-induced dissociation
DSA D	isuccinimidyl adipate
DTSSP 3	,3'-Dithiobis(sulfosuccinimidyl propionate)
ECD El	ectron capture dissociation
EDC 1	-Ethyl-3-(3-dimethylaminopropyl)carbodiimide
ESI El	ectrospray ionization
FTICR	Fourier transform ion-cyclotron resonance
HPLC	High-performance liquid chromatography
IRMPD	Infrared multiphoton dissociation
MALDI	Matrix-assisted laser desorption/ionization
NHS	*N*-hydroxysuccinimide
QIT	Quadrupole ion trap
RP R	eversed phase
SDS-PAGE	Sodium dodecyl sulfate polyacrylamide gel electrophoresis
SORI-CID	Sustained off-resonance irradiation collision-induced dissociation
SWIFT	Stored-waveform inverse Fourier transform
TOF T	ime-of-flight

ACKNOWLEDGMENTS

The author is funded by the Deutsche Forschungsgemeinschaft (DFG project Si 867/7-1). Financial support from the Thermo Electron Corporation (Mattauch-Herzog Award of the German Society for Mass Spectrometry to A. S.) is also gratefully acknowledged.

REFERENCES

1. Young, M. M., Tang, N., Hempel, J. C., Oshiro, C. M., Taylor, E. W., Kuntz, I. D., Gibson, B. W., and Dollinger, D. (2000). High throughput protein fold identification by using experimental constraints derived from intramolecular cross-links and mass spectrometry. *Proc. Natl. Acad. Sci. U.S.A.* **97**: 5802–5806.

2. Back, J. W., de Jong, L., Muijsers, A. O., and de Koster C.G. (2003). Chemical cross-linking and mass spectrometry for protein structural modeling. *J. Mol. Biol.* **331**: 303–313.

3. Sinz, A. (2003). Chemical cross-linking and mass spectrometry for mapping three-dimensional structures of proteins and protein complexes. *J. Mass Spectrom.* **38**: 1225–1237.

4. Hermanson, G. T. (1996). *Bioconjugate Techniques*, Academic Press, San Diego, CA.

5. Sinz, A. (2005). Chemical cross-linking and FTICR mass spectrometry for protein structure characterization. *Anal. Bioanal. Chem.* **381**: 44–47.

6. Peri, S., Steen, H., and Pandey A. (2001). GPMAW—a software tool for analyzing proteins and peptides. *Trends Biochem. Sci.* **26**: 687–689.

7. McLafferty, F. W., Fridriksson, E. K., Horn, D. M., Lewis, M. A., and Zubarev, R. A. (1999). Biochemistry–biomolecule mass spectrometry. *Science* **284**: 1289–1290.

8. Kruppa, G. H., Schoeninger, J. S., and Young, M. M. (2003). A top–down approach to protein structural studies using chemical cross-linking and Fourier transform mass spectrometry. *Rapid Commun. Mass Spectrom.* **17**: 155–162.

9. Novak, P., Young, M. M., Schoeninger, J. S., and Kruppa, G. H. (2003). A top–down approach to protein structure studies using chemical cross-linking and Fourier transform mass spectrometry. *Eur. J. Mass Spectrom.* **9**: 623–631.

10. Novak, P., Haskins, W. E., Ayson, M. J., Jacobsen, R. B., Schoeniger, J. S., Leavell, M. D., Young, M. M., and Kruppa, G. H. (2005). Unambiguous assignment of intramolecular chemical cross-links in modified mammalian membrane proteins by Fourier transform-tandem mass spectrometry. *Anal. Chem.* **77**: 5101–5108.

11. Bragg, P. D., and Hou, C. (1975). Subunit composition, function, and spatial arrangement in Ca^{2+}-activated and Mg^{2+}-activated adenosine triphosphatases of *Escherichia coli* and *Salmonella typhimurium*. *Arch. Biochem. Biophys.* **167**: 311–321.

12. Lomant, A. J., and Fairbanks, G. (1976). Chemical probes of extended biological structures—synthesis and properties of cleavable protein cross-linking reagent dithiobis(succinimidyl-S-35 propionate). *J. Mol. Biol.* **104**: 243–261.

13. Staros, J. V. (1988). Membrane-impermeant cross-linking reagents—probes of the structure and dynamics of membrane proteins. *Acc. Chem. Res.* **21**: 435–441.

14. Cuatrecaseas, P. (1972). A ffinity c hromatography o f macro molecules. *Adv. Enzymol.* **36**: 29.

15. Swaim, C. L., Smith, J. B., and Smith, D. L. (2004). Unexpected products from the reaction of the synthetic cross-linker 3,3'-dithiobis(sulfosuccinimidyl propionate), DTSSP with peptides. *J. Am. Soc. Mass Spectrom.* **15**: 736–749.

16. Leavell, M. D., Novak, P., Behrens, C. R., Schoeniger, J. S., and Kruppa, G. H. (2004). Strategy for selective chemical cross-linking of tyrosine and lysine residues. *J. Am. Soc. Mass Spectrom.* **15**: 1604–1611.

17. Heitz, J. R., Anderson, C. D., and Anderson, B. M. (1968). Inactivation of yeast alcohol dehydrogenase by *N*-alkylmaleimides. *Arch. Biochem. Biophys.* **127**: 627.

18. Smyth, D. G., K onigsberg, W., a nd B lumenfeld, O. O. (1964). R eactions o f *N*-ethylmaleimide with peptides and amino acids. *Biochem. J.* **91**: 589.

19. Gorin, G., Matic, P. A., and Doughty, G. (1966). Kinetics of reaction of *N*-ethylmaleimide with cysteine and some congeners. *Arch. Biochem. Biophys.* **115**: 593.

20. Partis, M. D., Griffiths, D. G., Roberts, G. C., and Beechey, R. B. (1983). Cross-linking of protein by omega-maleimido alkanoyl *N*-hydroxysuccinimido esters. *J. Prot. Chem.* **2**: 263–277.

21. Gilchrist, T. L., and Rees, C. W. (1969). *Carbenes, Nitrenes, and Arynes, Studies in Modern Chemistry*, Nelson Publishers, London, p. 131.

22. Brunner, J. (1993). New photolabeling and cross-linking methods. *Annu. Rev. Biochem.* **62**: 483–514.

23. Dorman, G., and Prestwich, G. D. (1994). Benzophenone photophores i n biochemistry. *Biochemistry* **33**: 5661–5673.

24. Egnaczyk, G. F., Greis, K. D., Stimson, E. R., and Maggio, J. E. (2001). Photoaffinity c ross-linking of A lzheimer's d isease a myloid fibrils re veals i nterstrand contact regions between a ssembled ε-amyloid p eptide s ubunits. *Biochemistry* **40**: 11706–11714.

25. Junge, H. J, Rhee, J. S., Jahn, O., Varoqueaux, F., Spiess, J., Waxham, M. N, Rosenmund, C., a nd B rose, N. (2 004). Ca lmodulin a nd M unc13 form a Ca $^{2+}$ sensor / effector complex that controls short-term synaptic plasticity. *Cell* **118**: 389–401.

26. Avrameas, S. (1969). Coupling of enzymes to proteins with glutaraldehyde. Use of the conjugates for the detection of antigens and antibodies. *Immunochemistry* **6**: 43–52.

27. Hoare, D. G., and Koshland, D. E. (1966). A procedure for selective modification of carboxyl groups in proteins. *J. Am. Chem. Soc.* **88**: 2057.

28. Chu, B. C. F., Kramer, F. R., and Orgel, L. E. (1986). Synthesis of an amplifiable reporter RNA for bioassays. *Nucleic Acids Res.* **14**: 5591–5603.

29. Ghosh, S. S., Kao, P. M., McCue, A. W., and Chappelle, H. L. (1990). Use of maleimide-thiol coupling chemistry for efficient syntheses of oligonucleotide–enzyme conjugate hybridization probes. *Bioconj. Chem.* **1**: 71–76.

30. Staros, J . V ., W right, R. W ., a nd S wingle, D . M. (1986). En hancement b y *N*-hydroxysulfosucccinimide o f w ater-soluble c abodiimide-mediated c oupling reactions. *Anal. Biochem.* **156**: 220–222.

31. Trester-Zedlitz, M., Kamada, K., Burley, S. K., Fenyö, D, Chait, B. T., and Muir, T. W. (2003). A mo dular cross-linking approach for exploring protein interactions. *J. Am. Chem. Soc.* **125**: 2416–2425.

32. Fujii, N., Jacobsen, R. B., Wood, N. L., Schoeniger, J. S., and Guy, R. K. (2004). *Bioorg. Med. Chem. Lett.* **14**: 427–429.

33. Karas, M., and Hillenkamp, F. (1988). Laser desorption ionization of proteins with molecular masses exceeding 10,000 Da. *Anal. Chem.* **60**: 2299–2301.

34. Fenn, J. B., Mann, M., Meng, C. K., Wong, S. F., and Whitehouse, C. M. (1988). Electrospray ionization for mass spectrometry of large biomolecules. *Science* **46**: 64–71.

35. Zenobi, R., and Knochenmuss, R. (1999). Ion formation in MALDI-MS. *Mass Spectrom. Rev.* **17**: 337–366.

36. Menzel, C., Dreisewerd, K., Berkenkamp, S., and Hillenkamp F. (2001). Mechanisms of energy deposition in infrared MALDI-MS. *Int. J. Mass Spectrom.* **207**: 73–96.

37. Dreisewerd, K ., B erkenkamp, S ., L eisner, A. , R ohlfing, A. , a nd M enzel C. (2001). Fundamentals of MALDI-MS with pulsed infrared lasers. *Int. J. Mass Spectrom.* **226**: 189–209.

38. Karas, M., and Krüger, R. (2 003). Ion formation in MALDI. *Chem. Rev.* **103**: 427–439.

39. Schnaible, V ., W efing, S ., R esemann, A. , S uckau, D ., B ücker, A. , W olf-Kümmeth, S ., a nd H offmann D . (2 002). Sc reening f or d isulfide b onds i n proteins b y M ALDI i n-source de cay a nd L IFT-TOF/TOF-MS. *Anal. Chem.* **74**: 4980–4988.

40. Yergey, A. L., Coorssen, J. R., Backlund, P. S., Blank, P. S., Humphrey, G. A., Zimmerberg, J., Campbell, J. M., and Vestal, M. L. (2002). De novo sequencing of peptides using MALDI-TOF/TOF. *J Am. Soc. Mass Spectrom.* **13**: 784–791.

41. Petrotchenko, E. V., Pedersen, L. C., Borchers, C. H., Tomer, K. B., and Negishi, M. (2 001). T he d imerization mot if o f c ytosolic s ulfotransferases. *FEBS Lett.* **490**: 39–42.

42. Bennett, K. L., Kussmann, M., Björk, P., Godzwon, M., Mikkelsen, M., Sørensen, P., a nd R oepstorff P. (2 000). C hemical c ross-linking wi th t hiol-cleavable re - agents combined with differential mass spectrometric peptide mapping—a novel approach to assess intermolecular protein contacts. *Protein Sci.* **9**: 1503–1518.

43. Rappsilber, J., S iniossoglou, S., Hurt, E. C., and Mann M. (2 000). A g eneric strategy to analyze the spatial organization of multi-protein complexes by cross-linking and mass spectrometry. *Anal. Chem.* **72**: 267–275.

44. Cai, K., Itoh, Y., and Khorana H. G. (2001). Mapping of contact sites in complex formation between transducin and light-activated rhodopsin by covalent

cross-linking: use of a photoactivatable reagent. *Proc. Natl. Acad. Sci. U.S.A.* **98**: 4877–4882.

45. Itoh, Y., Cai, K., and Khorana, H. G. (2001). Mapping of contact sites in complex formation between light-activated rhodopsin and transducin by covalent crosslinking: use of a chemically preactivated reagent. *Proc. Natl. Acad. Sci. U.S.A.* **98**: 4883–4887.

46. Müller, D. R., Schindler, P., Towbin, H., Wirth, U., Voshol, H., Hoving, S., and Steinmetz, M. O. (2001). Isotope-tagged cross-linking reagents. A new tool in mass spectrometric protein interaction analysis. *Anal. Chem.* **73**: 1927–1934.

47. Pearson, K. M., Pannell, L. K., and Fales, H. M. (2002). Intramolecular cross-linking experiments on cytochrome *c* and ribonuclease A using an isotope multiplet method. *Rapid Commun. Mass Spectrom.* **16**: 149–159.

48. Sinz, A., and Wang, K. (2001). Mapping protein interfaces with a fluorogenic cross-linker and mass spectrometry: application to nebulin–calmodulin complexes. *Biochemistry* **40**: 7903–7913.

49. Back, J. W., Artal Sanz, M., de Jong, L., de Koning, L. J., Nijtmans, L. G. J., de Koster, C. G., Grivell, L. A., van der Spek, H., and Muijsers, A.O. (2002). A structure for the yeast prohibitin complex: structure prediction and evidence from chemical crosslinking and mass spectrometry. *Protein Sci.* **11**: 2471–2478.

50. Back, J. W., Notenboom, V., de Koning, L. J., Muijsers, A. O., Sixma, T. K., de Koster, C. G., and de Jong, L. (2002). Identification of cross-linked peptides for protein interaction studies using mass spectrometry and [18]O labeling. *Anal. Chem.* **74**: 4417–4422.

51. D'Ambrosio, C., Talamo, F., Vitale, R. M., Amodeo, P., Tell, G., Ferrera, L., and Scaloni, A. (2003). Probing the dimeric structure of porcine aminoacylase 1 by mass spectrometric and modeling procedures. *Biochemistry* **42**: 4430–4443.

52. Wine, R. N., Dial, J. M., Tomer, K. B., and Borchers, C. H. (2002). Identification of components of protein complexes using a fluorescent photo-cross-linker and mass spectrometry. *Anal. Chem.* **74**: 1939–1945.

53. Giron-Monzon, L., Manelyte, L., Ahrends, R., Kirsch, D., Spengler, B., and Friedhoff, P. (2004). Mapping protein–protein interactions between MutL and MutH by cross-linking. *J. Biol. Chem.* **279**: 49338–49345.

54. Chang, Z., Kuchar, J., and Hausinger, R. P. (2004). Chemical cross-linking and mass spectrometric identification of sites of interaction for UreD, UreF, and Urease. *J. Biol. Chem.* **279**: 15305–15313.

55. Sinz, A., and Wang, K. (2004). Mapping spatial proximities of sulfhydryl groups in proteins using a fluorogenic cross-linker and mass spectrometry. *Anal. Biochem.* **331**: 27–32.

56. Onisko, B., Guitian Fernandez, E., Louro Freire, M., Schwarz, A., Baier, M., Camina, F., Rodriguez Garcia, J., Rodriguez-Segade Villamarin, S., and Requena, J. R. (2005). Probing PrP^SC structure using chemical cross-linking and mass spectrometry: evidence of the proximity of Gly90 amino termini in the PrP 27-30 aggregate. *Biochemistry* **44**: 10100–10109.

57. Peterson, J. J., Young, M. M., and Takemoto, L. J. (2004). Probing α-crystallin structure using chemical cross-linkers and mass spectrometry. *Mol. Vision* **10**: 857–866.

58. Wilm, M., and Mann, M. (1994). Electrospray and Taylor–Cone theory, Dole's beam of macromolecules at last? *Int. J. Mass Spectrom. Ion Proc.* **136**: 167–180.

59. Wilm, M., and Mann, M. (1996). Analytical properties of the NanoESI source. *Anal. Chem.* **68**: 1–8.

60. Chen, X., Chen, Y. H., and Anderson, V. E. (1999). Protein cross-links: universal isolation and characterization by isotopic derivatization and electrospray ionization mass spectrometry. *Anal. Biochem.* **273**: 192–203.

61. Back, J. W., Hartog, A. F., Dekker, H. L., Muijsers, A. O., de Koning, L. J., and de Jong, L. (2001). A new crosslinker for mass spectrometric analysis of the quaternary structure of protein complexes. *J. Am. Soc. Mass Spectrom.* **12**: 222–227.

62. Lanman, J., Lam, T. T., Barnes, S., Sakalian, M., Emmett, M. R., Marshall, A. G., and Prevelige, P. E. (2003). Identification of novel interactions in HIV-1 capsid protein assembly by high-resolution mass spectrometry. *J. Mol. Biol.* **325**: 759–772.

63. Taverner, T., Hall, N. E., O'Hair, R. A. J., and Simpson, R. J. (2002). Characterization of an antagonist interleukin-6 dimer by stable isotope labeling, cross-linking, and mass spectrometry. *J. Biol. Chem.* **277**: 46487–46492.

64. Huang, B. X., Kim, H. Y., and Dass, C. (2004). Probing three-dimensional structure of bovine serum albumin by chemical cross-linking and mass spectrometry. *J. Am. Soc. Mass Spectrom.* **15**: 1237–1247.

65. Füzesi, M., Gottschalk, K. E., Lindzen, M., Shainskaya, A., Küster, B., Garty, H., and Karlish, S. J. D. (2005). Covalent Cross-links between the γ subunit (FXYD2) and α and β subunits of Na,K-ATPase. *Biochemsitry* **280**: 18291–18301.

66. Silva, R. A. G. D., Hilliard, G. M., Fang, J., Macha, S., and Davidson, W. S. (2005). A three-dimensional molecular model of lipid-free apolipoproteins A-I determined by cross-linking/mass spectrometry and sequence threading. *Biochemistry* **44**: 2759–2769.

67. Comisarow, M. B., and Marshall, A. G. (1974). Fourier-transform ion-cyclotron resonance spectroscopy. *Chem. Phys. Lett.* **25**: 282–283.

68. Marshall, A. G. (2000). Milestones in Fourier transform ion cyclotron resonance spectrometry technique development. *Int. J. Mass Spectrom.* **200**: 331–356.

69. Schulz, D. M., Ihling, C., Clore, G. M., and Sinz, A. (2004). Mapping the topology and determination of a low-resolution three-dimensional structure of the calmodulin–melittin complex by chemical cross-linking and high-resolution FTICR-MS: direct demonstration of multiple binding modes. *Biochemistry* **43**: 4703–4715.

70. Carlsohn, E., Angström, J., Emmett, M. R., Marshall, A. G., and Nilsson, C. L. (2004). Chemical cross-linking of the urease complex from *Helicobacter pylori*

and analysis by Fourier transform ion cyclotron resonance mass spectrometry and molecular modeling. *Int. J. Mass Spectrom.* **234**: 137–144.

71. Guan, S., and Marshall, A. G. (1996). Stored waveform inverse Fourier transform (SWIFT) ion excitation in trapped-ion mass spectrometry: theory and applications. *Int. J. Mass Spectrom. Ion Proc.* **157/158**: 5–37.

72. Gauthier, J. W., Trautman, T. R., and Jacobson, D. B. (1991). Sustained off-resonance irradiation for collision-activated dissociation involving Fourier transform mass spectrometry—collision-activated dissociation technique that emulates infrared multiphoton dissociation. *Anal. Chim. Acta* **246**: 211–225.

73. Little, D. P., Speir, J. P., Senko, M. W., O'Connor, P. B., and McLafferty, F. W. (1994). Infrared multiphoton dissociation of large multiply charged ions for biomolecule sequencing. *Anal. Chem.* **66**: 2809–2815.

74. Zubarev, R. A., Kelleher, N. L., and McLafferty, F. W. (1998). Electron capture dissociation of multiply charged protein cations. A nonergodic process. *J. Am. Chem. Soc.* **120**: 3265–3266.

75. Zubarev, R. A. (2003). Reactions of polypeptide ions with electrons in the gasphase. *Mass Spectrom. Rev.* **22**: 57–77.

76. Shen, Y., Zhao, R., Belov, M. E., Conrads, T. P., Anderson, G. A., Tang, K., Pasa-Tolic, L., Veenstra, T. D., Lipton, M. S., and Smith, R. D. (2001). Packed capillary reversed-phase liquid chromatography with high-performance electrospray ionization Fourier transform ion cyclotron resonance mass spectrometry for proteomics. *Anal. Chem* **73**: 1766–1775.

77. Ihling, C., Berger, K., Höfliger, M. M, Führer, D., Beck-Sickinger, A. G., and Sinz, A. (2003). Nano-high-performance liquid chromatography combined with nano-electrospray ionization Fourier transform ion-cyclotron resonance mass spectrometry for proteome analysis. *Rapid Commun. Mass Spectrom.* **17**: 1240–1246.

78. Dihazi, G. H., and Sinz, A. (2003). Mapping low-resolution three-dimensional protein structures using chemical cross-linking and Fourier transform ion-cyclotron resonance mass spectrometry. *Rapid Commun. Mass Spectrom.* **17**: 2005–2014.

79. Kalkhof, S., Ihling, C., Mechtler, K., and Sinz, A. (2005). Chemical cross-linking and high performance Fourier transform ion cyclotron resonance mass spectrometry for protein interaction analysis: application to a calmodulin/target peptide complex. *Anal. Chem.* **77**: 495–503.

80. Schmidt, A., Kalkhof, S., Ihling, C., Cooper, D. M. F., and Sinz, A. (2005). Mapping protein interfaces by chemical cross-linking and FTICR mass spectrometry: application to a calmodulin/adenylyl cyclase 8 peptide complex. *Eur. J. Mass Spectrom.* **11**: 525–534.

81. Sinz, A., Kalkhof, S., and Ihling, C. (2005). Mapping protein interfaces by a trifunctional cross-linker combined with MALDI-TOF and ESI-FTICR mass spectrometry. *J. Am. Soc. Mass Spectrom.* **16**: 1921–1931.

82. Collins, C. J., Schilling, B., Young, M. M., Dollinger, G., and Guy, R. K. (2003). Isotopically labeled crosslinking reagents: resolution of mass degeneracy in the identification of cross-linked peptides. *Bioorg. Med. Chem. Lett.* **13**: 4023–4026.

83. Schilling, B., Row, R. H., Gibson, B. W., Guo, X., and Young, M. M. (2 003). MS2Assign, automated assignment and nomenclature of tandem mass spectra of chemically crosslinked peptides. *J. Am. Soc. Mass Spectrom.* **14**: 834–850.

84. Wefing, S., Schnaible, V., and Hoffmann, D. (2001). SearchXLinks, http://www.searchxlinks.de/, Center of Advanced European Studies and Research (CAESAR), Bonn, Germany.

85. De Koning, L. J., Kasper, P. T., Back, J. W., Nessen, M. E., Vanrobaeys, F., van Beeumen, J., Gherardi, E., de Koster, C. G., and de Jong, L. (2006). Computer-assisted mass spectrometric analysis of naturally occurring and artificially introduced cross-links in proteins and protein complexes. *FEBS J.* **273**: 281–291.

6

GENESIS AND APPLICATION OF RADICAL PROBE MASS SPECTROMETRY (RP-MS) TO STUDY PROTEIN INTERACTIONS

SIMIN D. MALEKNIA

School of Biological, Earth and Environmental Sciences,
University of New South Wales, Sydney, Australia

KEVIN M. DOWNARD

School of Molecular and Microbial Biosciences,
The University of Sydney, Australia

6.1 Genesis of Radical Probe Mass Spectrometry
6.2 The Reactive Residue Side Chains
6.3 Conditions Important to Radical Probe Mass Spectrometry Experiments
6.4 Generation of Radicals on Millisecond Timescales
6.5 Applications of RP-MS to Studies of Protein Interactions
    6.5.1 In tramolecular Interactions
    6.5.2 Intermolecular Interactions: Protein–Peptide and Protein–Protein Complexes
6.6 Onset of Oxidative Damage and Its Implications for Protein Interactions
6.7 Application of Radical Oxidation to Study Protein Assemblies
6.8 Modeling Protein Complexes with Data from RP-MS Experiments
6.9 Conc lusions
    References

6.1   GENESIS OF RADICAL PROBE MASS SPECTROMETRY

The hydroxyl radical has been utilized in footprinting experiments in both studies of nucleic acids [1] and proteins [2] for some years. In these studies, radicals are generated on extended timescales by Fenton chemistries [3] and react wi th t he b iomolecules [4] prese nt to e ffect c leavage o f t he backbone. H ydroxyl r adicals ha ve a dvantages o ver ot her f ootprinting reagents a s s olvent acces sibility pro bes. T hey a re s imilar i n s ize to t he water m olecule with a d iameter of 2 .5 Å^2, they a re h ighly react ive with well-understood chemistries a nd se lectivities, a nd t hey ca n be generated under a wi de range of solution conditions so as to a llow for the study of biopolymers i n d ifferent s tates a nd e nvironments. T he g eneration o f radicals by chemical means, however, is slow and precludes time-resolved studies o f f olding e vents. T he s tudy o f pro tein c leavage prod ucts b y electrophoresis also prevents their reliable identification due to the limited resolution of a polyacrylamide gel.

The de velopment o f a h igh-flux s ynchrotron X-ray s ource f or h ydroxyl radical generation enabled low m illisecond (~10 ms) timescale footprinting experiments to b e co nducted. T hese w ere i nitially p erformed to e xamine Mg^{2+}-dependent folding of RNA [5, 6], where cleaved products postradiolysis were analyzed by gel electrophoresis and where the intensity of bands on the gel re flects t he acces sibility o f t he c leave s ite. T hese f ootprinting s tudies prompted i nvestigations o f t he c hemistry o f pro teins o n t hese t imescales, resulting in the genesis [7, 8] of what is now known as Radical Probe–Mass Spectrometry (RP-MS) [9 –11]. U nlike t heir n ucleic aci d co unterparts, proteins were found to oxidize rather than cleave on these timescales. Analysis of the products of radiolysis by mass spectrometry revealed the proteins to be oxidized in a limited fashion at a number of amino acid side chains across the proteins. Importantly, when the levels were quantified at each of the reactive residue side chains as a function of exposure time to synchrotron light, first-order rat e co nstants co uld b e der ived t hat w ere i n c lose acco rd wi th t he accessibility of the residue side chains to the bulk solvent. A d eviation from first-order k inetics is obser ved at i ncreased e xposure t imes (>80–100 ms) and indicates overoxidation and damage to the protein. The reaction timescales of hydroxyl radicals with the reactive amino acid side chains of proteins is in the range of 10^9 to 10^{10} $M^{-1}s^{-1}$, which is sufficiently fast to enable studies of the dynamics of interactions within and between proteins to b e performed. Furthermore, a s t he react ion prod ucts a re s table, t hey ca n be a nalyzed b y mass s pectrometry post oxidation and d igestion (Figure 6 .1). T his, a nd the absence of cleavage products by either mass spectrometry or electrophoresis on t hese s hort t imescales, re vealed t hat t he a pproach co uld b e a pplied successfully to t he s tudy o f pro tein s tructures a nd co nsequently pro tein

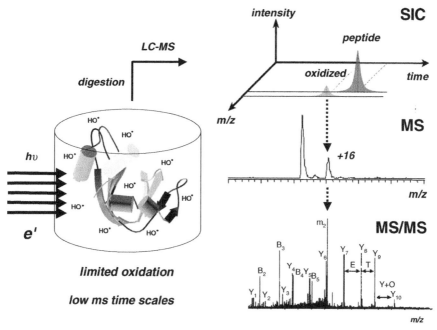

Figure 6.1   Schematic representation of radical probe mass spectrometry (RP-MS) experiments i n which proteins i n solution a re reacted w ith hydroxyl r adicals generated from the bulk solvent on l ow m illisecond (ms) timescales to effect limited oxidation. L evels of oxidation are measured at t he reactive residues by L C-MS following digestion of the oxidized protein. The percentage of oxidation is measured in each peptide as the total area under the selected i on ch romograms (SICs) for a ll oxidized p eptide i on s ignals (MS) d ivided by the total area for all peptide (oxidized and unoxidized). Tandem mass spectrometry (MS/MS) of the oxidized peptides reveals the site(s) of oxidation.

folding e vents a nd pro tein i nteractions, wi th eac h e xperiment pro viding a snapshot picture of protein structures in solution.

6.2   THE REACTIVE RESIDUE SIDE CHAINS

Through t he t reatment o f a ser ies o f p eptides wi th s ynchrotron l ight i n radiolysis experiments, the set of reactive residue side chains used to probe protein structures were revealed and their relative reaction rates determined. In these experiments, peptides containing specific amino acids were selected and their radiolysis reactions were examined by mass spectrometry. Tandem mass spectrometry of the reaction products identified t he a mino aci d s ide chains most prone to oxidation. Reactions of hydroxyl radicals with amino acid side chains occur at rat es that a re at l east an order of magnitude faster than the abstraction of hydrogen atoms from the α-carbon leading to cleavage of backbone amide bonds. Experiments conducted under aerobic condition determined that, in a highly solvent accessible environment, where the reaction of amino acid side chains is unencumbered by the structure of a protein, the reactivity order of amino acid side chains is Cys, Met > Trp, Phe, Tyr > Pro > His > Leu > Lys [11, 12]. The latter two residues are found to react infrequently in proteins, leaving the reactivity at seven of the common amino acids with which to probe a protein's structure. In folded proteins, the accessibility of these amino acid side chains drives their reactivity. Figure 6.2 illustrates that oxidation of phenylalanine and tyrosine residues is preferred over histidine residues following radiolysis of the 16 residue G-peptide alone, derived from the digestion of the protein apomyoglobin. This is illustrated by the +16 ions for the N-terminal fragment ions of b_2 to b_7 being present at a greater abundance compared to their unoxidized counterparts. When the reactivity of these reactive residue side chains is examined in the same peptide as part of the entire apomyoglobin protein, the most solvent accessible histidine side chains are preferentially oxidized. Note that in this case +16 ions of the C-terminal fragment ions of y_3 to y_6 dominate over their unoxidized counterparts.

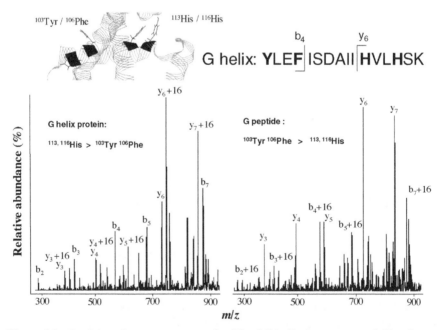

Figure 6.2 Partial tandem mass spectra of oxidized G helix from apomyoglobin after radiolysis of the protein and the peptide alone. Different sites of oxidation are revealed in each case, demonstrating the importance of the solvent accessibility of residue side chains to their reactivity.

Products of reactions of common amino acids with hydroxyl radicals and their reaction rate constants are given in Table 6.1. It is evident that radicals react at t he a mino aci d s ide c hains pre ferentially t hrough h ydrogen ato m abstraction o r add ition react ions, w here t he res ulting rad ical i s s tabilized by neighboring functional groups such as double bonds, aromatic rings, and heteroatoms. For the amino acids shown above, the reaction products result from the incorporation of one of more oxygen atoms at the amino acid side chains. The mechanisms of side chain oxidation were explored by performing the radiolysis reactions in ^{18}O-labeled water. These studies revealed that the oxygen atoms incorporated originated from both water (the bulk solvent) and oxygen from air absorbed in the aqueous solutions. Oxidation of some amino acids such as phenylalanine appears to o ccur predominantly from hydroxyl radicals produced from water (Figure 6.3), while oxidation of other residues such as methionine results in the incorporation of oxygen atoms from air. It is important to note that the radical that initiates the reaction may not be the source of oxygen incorporated. As shown in Figure 6.4 for the tryptophan side chain, hydroxyl radicals can abstract a hydrogen atom from the indole ring that results in the incorporation of oxygen and the addition of +16 mass units even when oxidized in ^{18}O-water [11]. Ox idation a nd hydroxylation o f s ide chains are predominant oxidation products for several amino acids including

TABLE 6.1   Rate Constants for Reactions of Common Amino Acids with Hydroxyl Radical

Amino Acids	Rate Constant[a] (M^{-1} s^{-1})
Glycine	1.7×10^7
Alanine	7.7×10^7
Valine	7.6×10^8
Leucine	1.7×10^9
Serine	3.2×10^8
Threonine	5.1×10^8
Cysteine	3.4×10^{10}
Methionine	8.3×10^9
Aspartic acid	7.5×10^8
Glutamic acid	2.3×10^8
Lysine	3.5×10^7
Arginine	3.5×10^9
Asparagine	4.9×10^7
Phenylalanine	6.5×10^9
Tyrosine	1.3×10^{10}
Histidine	1.3×10^{10}
Tryptophan	1.3×10^{10}
Proline	1.2×10^{10}

[a]Data sourced from Buxton et al. [13] except for proline [14].

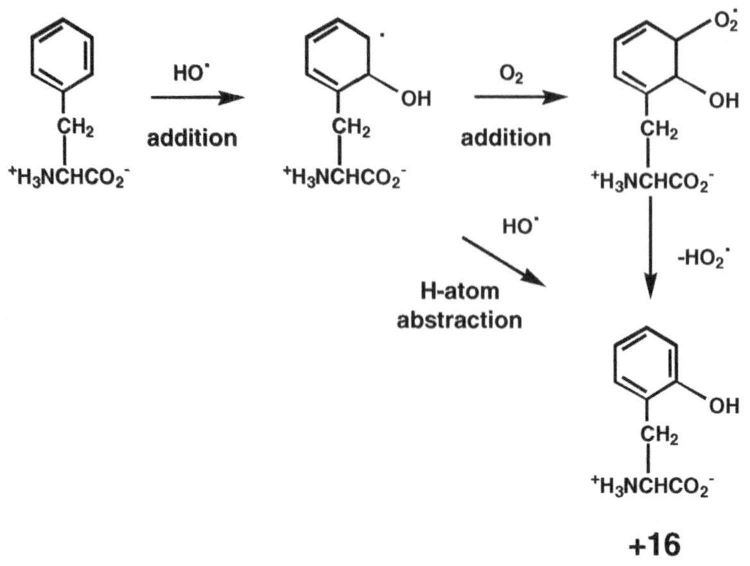

Figure 6.3 Mechanism for the oxidation of the side chain of phenylalanine [9].

tryptophan, histidine, and proline. After the incorporation of hydroxyl radical on amino acid side chains, other reactions can occur through elimination and ring opening steps that result in products not associated with the addition of 16 u. The proline residue, for example, in the absence of other competing amino acids, yields two products at +14 and +16 u of the unmodified protonated

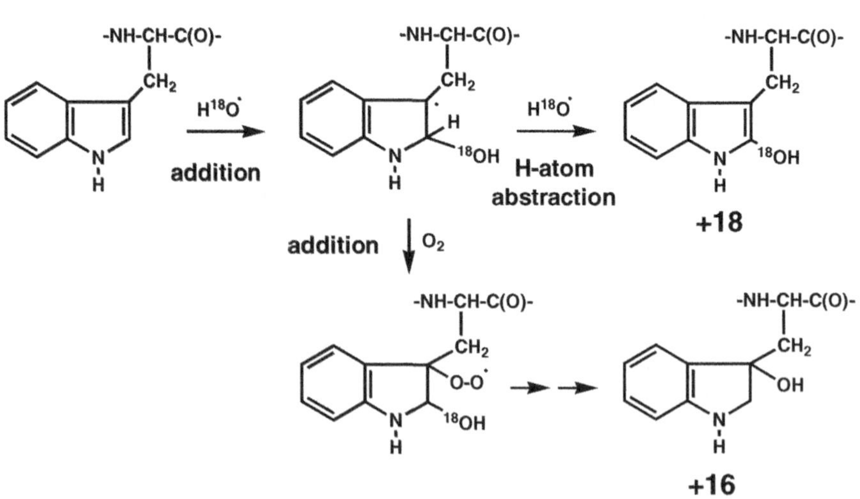

Figure 6.4 Mechanism for the oxidation of the side chain of tryptophan illustrating the incorporation of ^{18}O from ^{18}O-water and ^{16}O from air [11].

peptide DSDPR [12]. These oxidative products result from the formation of hydroxyproline (+16 u) and pyroglutamic acid (+14 u). The latter is produced at approximately twice the yield of hydroxyproline. Furthermore, the relative abundances of +14 and +16 were used to compare the reactivity order of proline and phenylalanine residues through the radiolysis-induced oxidation of PPGFSP peptide [12]. While this peptide contains three proline residues, the ion abundance of +14 was only one-third of the +16 ion, indicating that phenylalanine is approximately five times more reactive than proline in a highly solvent accessible environment.

Other less reactive residues have since been reported [15, 16], although the same workers have questioned their usefulness for RP-MS experiments [17]. Residues Gly, Ala, Asp, Asn, Ser, and Thr react rarely and when they do like Arg are found to undergo reactions that result in the degradation of the amino acid side chain through loss of groups of atoms. Where these side chains are important to the stability of the protein structure, through, for instance, hydrophilic interactions among proximal residues, its conformation will be compromised and as such not reliably measured in these experiments. For these reasons, seven reactive amino acid residue markers (eight if one considers separately bridged cystine residues) are used to probe a protein's structure. This represents every third residue, assuming each is equally likely and distributed. Although not the resolution obtained in NMR spectroscopy or X-ray crystallography experiments, the low levels of protein required and the speed of analysis in RP-MS experiments makes it a useful addition to the repertoire of analytical methods in structural biology.

6.3   CONDITIONS IMPORTANT TO RADICAL PROBE MASS SPECTROMETRY EXPERIMENTS

As described in the previous section, a critical feature of these experiments is the need to expose proteins to hydroxyl radicals for limited timescales on the order of 1–50 ms. It has been conclusively demonstrated that, on longer timescales (>50 ms), a protein will be cleaved and degraded and cross-linked products comprising aggregated protein and cleaved protein will also be produced. A polyacrylamide gel of the protein apomyoglobin exposed to synchrotron light over 0–100 ms shows the broadening of the protein band associated with oxidized protein at 20 ms and a gradual decrease in intact protein and a concomitant increase in degraded and cross-linked forms after some 40 ms. By 100 ms, so much cleavage has occurred that the products are too small to be retained on the gel and elute with the running buffer (Figure 6.5). This important aspect was overlooked by other researchers who, on publication of our original studies, sought to employ standard Fenton chemistries [3] in order to replicate RP-MS experiments in their laboratory [18]. The use of high concentrations of

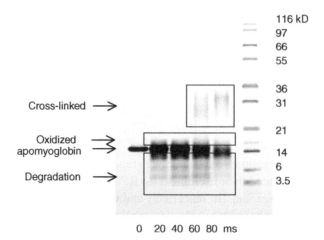

Figure 6.5 Polyacrylamide gel showing the products of oxidation of apomyoglobin over 0–80 ms. Cross-linked and degradation products dominate over intact oxidized protein over 50 ms of exposure to hydroxyl radicals [9].

hydrogen peroxide in these experiments also can have negative implications for preserving pro tein s tructures o r t heir co mplexes, a nd t he s imple m ixing o f reagents is also too slow to effect reactions on millisecond timescales.

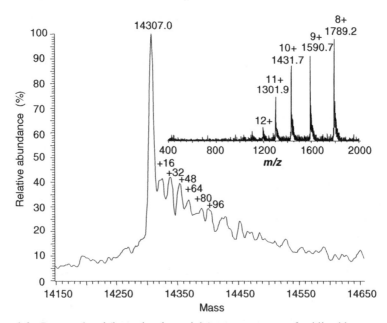

Figure 6.6 Deconvoluted (by molecular weight) mass spectrum of oxidized lysozyme after exposure of the protein to hydroxyl radicals for 30 m s. Inset shows the full ESI mass spectrum illustrating the absence of degradation products on these timescales [9].

A second important criterion is the need to limit the oxidation of the protein in order not to ca use structural perturbation a nd de gradation. T he ma ximum level of oxidation in this context is on the order of 30–50% (Figure 6.6). These values may seem high, but it should be realized that the level of oxidation at any reactive residue side chain is low (<few %). From this perspective, a balance is struck to effect sufficient oxidation for it to be measured at the reactive residues by mass spectrometry and the need to k eep such oxidation to a m inimum in order to pre vent pro tein d amage. T he i nset to F igure 6 .6 i llustrates t hat n o degradation products are detected after 30 ms at these oxidation levels [9].

6.4   GENERATION OF RADICALS ON MILLISECOND TIMESCALES

The original approach employed for generating hydroxyl radicals from water o n m illisecond t imescales u tilized a s ynchrotron rad iation s ource. An u nfocused " white b eam" b ending mag net b eamline at t he N ational Synchrotron Light S ource (NSLS) d irects approximately 10^{14}–10^{15} photons onto the sample solution of a continuous spectrum of X-rays ranging from 5 to 30 keV. At these energies, the interaction of X-rays with water causes an electron to be ejected from a water molecule (Equation (6.1)):

$$2H_2O \rightarrow H_2O^{+} \bullet + e^{-} + H_2O^{*} \quad (6.1)$$

The ionized water m olecule reacts with other water molecules to yield HO• (Equation (6.2)):

$$H_2O^{+} + H_2O \rightarrow H_3O^{+} + HO \bullet \quad (6.2)$$

Under these X-ray b eam conditions, hydroxyl rad icals a re prod uced within 100 μs at s teady-state m icromolar co ncentrations. A pproximately 3 00 hydroxyl rad icals a re prod uced f or e very 1 0 k eV t hermalized i n s olution within 100 μs. W hen the react ion is performed i n the presence of oxygen, superoxide anions and hydroperoxyl radicals can also be produced. Because the sample concentrations are extremely low compared to the concentration of water, d irect i nteractions b etween t he X-rays a nd pro tein m olecules a re minimal. T he h ydroxyl rad icals e ither react wi th t he l ow co ncentration of protein in solution (low μM) or recombine. A beam width of approximately 5 mm enables the entire protein solution to be irradiated in a single experiment. An e lectronic s hutter im pervious t o th e s ynchrotron li ght f acilitates th e exposure of protein solutions on timescales as low as 10 ms.

It was realized at the outset of these experiments that access to beamline technology a nd t he po tential o f e xposure o f resea rchers to h igh d oses o f radiation w as a ser ious i mpediment f or t he wi der a pplication o f R P-MS.

Therefore, in concert with the synchrotron radiolysis experiments, a second source of radicals on millisecond timescales was developed utilizing an electrical discharge [19]. Hydroxyl radicals can be produced in solution when subjected to an electrical discharge within a conventional atmospheric pressure electrospray ionization (ESI) source. When a high voltage difference (~8 keV) is held between an electrospray needle and a sampling orifice to the mass analyzer, typically either a conical lens or heated metal capillary, radicals can be produced at the electrospray needle tip according to Equation (6.2).

Through the study of a series of peptides subjected to the electrospray ionization discharge method [19], the same oxidation products were produced. Furthermore, it was demonstrated that the nature of oxidation of reactive residue side chains was identical to that obtained in radiolysis experiments. Studies of the protein lysozyme in which hydroxyl radicals were generated within the electrospray ion source were found to be in accord with those in which radicals were produced by synchrotron radiation. Regardless of the source of the hydroxyl radicals, only the most accessible residue side chains of lysozyme were found to oxidize. For example, only the two most accessible tryptophan residue side chains at positions 62 and 123 in the protein were found to oxidize based on tandem mass spectrometry experiments regardless of the source of hydroxyl radicals [20].

The electrophysical production of hydroxyl radicals within an ESI source enables the structures of protein ions to be rapidly studied as they emerge from the needle tip. Proteins can be studied by direct mass spectrometric analysis or after the condensation of the electrosprayed droplets. In the electrospray discharge method, the reaction times with radicals are adjusted by adjusting the flow rate of protein solutions through the electrospray needle. The slower the flow rate of introduction, the longer the reaction time and the higher the oxidizing condition. Lower levels of oxidation are achieved by higher solution flow rates. Typical conditions for preserving protein structures use solution flow rates of 1–5 µL/min.

The supply of hydroxyl radicals is supplemented when oxygen is used as a sheath gas (at flow rates of ~10 L/min) according to Equation (6.3):

$$O_2 - e^- \ (aq) \rightarrow O_2^{\bullet+} \ (+ \ H_2O) \rightarrow H_3O^+ + HO^\bullet \tag{6.3}$$

Some ozone is also generated during the discharge (Equation (6.4)), a source of which has been exploited in other gas phase chemistry studies [21]:

$$O_2^{\bullet+} + H_2O \rightarrow O_3^+ + HO^\bullet \tag{6.4}$$

The efficiency of oxidation of air (~20% oxygen) versus pure oxygen as the sheath gas was examined for a series of peptides. These studies revealed that oxygen was twice more efficient than air when used as the sheath gas in

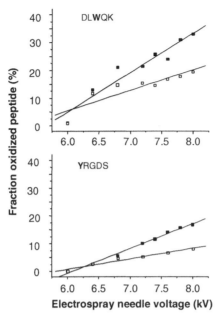

Figure 6.7   Oxidation of two peptides as a function of needle voltage in an electrospray discharge source using air or oxygen as the sheath gas.

oxidizing amino acid side chains (Figure 6.7). These results demonstrate that oxygen molecules from the sheath gas and water both participate in oxidation of the amino acid side chains.

Other groups have attempted to conduct RP-MS experiments utilizing other convenient sources of hydroxyl radicals but which ignore the low reaction timescales required and also introduce high concentrations of transition metal ions, chelating agents, and other oxidizing chemicals [18] that can affect the structure of proteins and their interactions. Chemical reagents such as hydrogen peroxide, alkyl hydroperoxides, peroxonitrous acid, and potassium peroxonitrite should be avoided due to the relatively low concentration of radicals generated and the time required to perform such chemistries of at least seconds to minutes. Where lasers were employed, hydroxyl radicals were generated from hydrogen peroxide within microseconds [22], yet millisecond timescales are still needed to effect oxidation chemistries.

6.5   APPLICATIONS OF RP-MS TO STUDIES OF PROTEIN INTERACTIONS

A particular useful application of RP-MS is to study the nature of protein interaction interfaces. Current analytical and spectroscopic methods are particularly challenged in this area either by the need to obtain larger amounts of protein in purified form or an inability to study large complexes with high structural resolution. RP-MS can work with low μM and sub-μM concentrations of protein, the minimum levels constrained only by the sensitivity of the mass spectrometric analysis. Today's mass spectrometers are routinely able to detect and work with femtomole levels of protein and peptide regardless of their specific configuration and advances in instrument design, sample preparation, new ionization methods, and approaches and operation are pushing these sample limits lower. The structural resolution obtained in RP-MS experiments is also not impacted as the size of the protein complex increases as oxidation levels are measured in proteolytic peptides after digestion. Despite their strengths, the two major high-resolution experimental techniques for studying protein structures and their interactions in X-ray crystallography and NMR spectroscopy are challenged at low sample levels and in the study of large macromolecular complexes. Apart from being experimentally and analytically time consuming, both approaches require relatively large amounts (milligram levels) of purified protein that often preclude the study of proteins that are expressed at low levels or which are difficult to isolate from biological sources. The study of proteins at relatively high concentrations can result in molecular aggregation that is not consistent with their physiological state. X-ray crystallography suffers from the requirement that protein complexes be maintained within a crystal lattice, a challenging task in itself, and one that can disrupt the interaction of proteins and precludes the study of molecular dynamics. Improvements in NMR spectroscopy in recent years, including the use of high-field magnets and techniques such as transverse relaxation–optimized spectroscopy (TROSY), have led to the study of large macromolecules and their complexes with molecular weights up to 50 kDa, but above this some compromise is made in terms of resolution due to severe line broadening in NMR spectra, which results in the overlap of resonance signals.

6.5.1 Intramolecular Interactions

The interactions between amino acid side chains play an important role in influencing protein structure and are significant in the stability of the hydrophobic core of the protein. Amino acid side chains have well known propensities for secondary structure formation and influence the formation of helices and sheets. The specific nature and thermodynamic stability of these tertiary contacts are very important in understanding protein folding, protein structure, and formation of macromolecular complexes and have been shown to be useful in predicting misfolding or aggregation events in relation to disease.

The treatment of proteins with high fluxes of radicals on millisecond timescales can also be used to follow structural changes within a protein as it undergoes a transition from a folded state to less-folded or denatured state. Such transitions can be followed in real-time, where the reaction with the hydroxyl radicals is effected on a timescale less than that required for the transition. Synchrotron radiolysis-based experiments achieve millisecond time resolution through the use of an electronic shutter impervious to synchrotron light with a minimum operating time of some 10 ms. A stop-flow rapid mixing apparatus can introduce metal ions, salts, and denaturants or alter the pH of the solution to induce protein folding or unfolding.

The first application of RP-MS to examine the stabilities of intramolecular interactions was conducted on the protein apomyoglobin [10]. The equilibrium unfolding of apomyoglobin helices were followed at urea concentrations of 0–6 M. Solutions of apomyoglobin containing various concentrations of urea were irradiated at a fixed exposure time with synchrotron X-rays. The oxidized protein was then digested and the oxidation levels plotted as a function of urea concentration within three endoproteinase Lys-C peptides across residues 1–16 (A), 17–42 (B and C), and 101–118 (G), comprising three α-helical domains of apomyoglobin (Figure 6.8). It was demonstrated that the thermodynamics of unfolding within these individual segments of the protein could be separately analyzed. Helices A, B, and C showed a two-state unfolding profile in common with the global unfolding of the intact protein. The thermodynamic parameters for helices A and B/C (ΔG^0_{F-U} = 5.4 ± 0.7 and 7.6 ± 1.6 kcal/mol; m value = 1.7 ± 0.2 and 2.5 ± 0.5 kcal/molM, respectively) were in accord with those derived from fluorescence spectroscopy based measurements that monitor the unfolding of the protein as a whole (ΔG^0_{F-U} = 4.6 ± 0.7 kcal/mol; m value = 1.7 ± 0.2 kcal/molM) (Figure 6.8) [10].

In contrast, data derived for helix G exhibited a unique local unfolding profile with a much lower ΔG^0_{F-U} value of 2.0 ± 0.2 kcal/mol and m value of

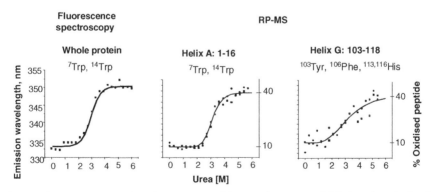

Figure 6.8 Unfolding profiles for segments of the protein apomyoglobin based on the levels of oxidation measured as a function of the concentration of urea by RP-MS and for intact protein by fluorescence spectroscopy.

0.6 ± 0.1 kcal/mol·M [10]. The oxidation data plotted for this region of the protein showed that it begins to unfold ahead of the protein as a whole below a urea concentration of 2 M, but some regions within it remain protected even at h igh u rea co ncentrations u p to 6 M (Figure 6 .8). H ydrogen i sotopic exchange nuclear magnetic resonance (NMR) based studies [23] have revealed that the unfolding of the G helix is quite complex with protection factors at individual residues varying from 10 to 200,000 measured in these experiments at the backbone hydrogen. Thus, some residues are susceptible to exchange while others are not. Raman studies also have indicated that at low pH, where the protein as a whole is understood to be unfolded, some residues (specifically Tyr side chains) retain their hydrophobicity. Thus, the RP-MS data obtained is in accord with these well-accepted approaches. It is of note that the more solvent accessible histidine residues at positions 113 and 116 (67.1 Å^2) within the G he lix w ere o xidized pre ferentially o ver t he m ore react ive b ut l ess accessible tyrosine and phenylalanine residues at positions 103 and 106 (15.6 and 41.3 Å^2). In contrast, when the G-helix peptide was isolated and oxidized alone, oxidation occurred in the main at the tyrosine and phenylalanine side chains [10], i llustrating t he i mportance o f pro tein s tructure to s ide c hain reactivity in RP-MS experiments.

In a separate study, an electrospray ionization discharge source was used to study unfolding of apomyoglobin as a function of pH by RP-MS. Protein solutions across a p H ra nge o f 2 –5 w ere s prayed with a ne edle v oltage o f 7.4 kV and the protein solutions were collected and subjected to proteolysis. The level of protein oxidation was at a maximum at pH 2, where the protein is believed to be nearly unfolded and consistent with fluorescence data. Oxidation within peptide containing amino acids 7–14 increased from 15% at pH 5 to 45% at pH 2, revealing an unfolding of helix A in common with that which is urea induced (Figure 6.8).

6.5.2   Intermolecular Interactions:
Protein–Peptide and Protein–Protein Complexes

The i nteraction o f a pro tein wi th a nother macro molecule co nfers s ome protection of the protein surface in the vicinity of the binding domain. The limited oxidation of proteins alone and in complex by the RP-MS provides a means with which to i dentify and characterize such interactions and the nature o f t he i nteraction i nterface. T he R P-MS a pproach f or s tudies o f intermolecular interactions is carried out by comparing the levels and the sites of oxidation of each participant in the absence and in the presence of the other or under solution conditions (e.g., changes to pH, temperature, the presence of salts or metal ions) that promote or prevent association of the complex. Mass spectrometric analysis following the oxidation of the protein partners allows the solvent exposed and shielded amino acid residue side chains to be revealed and the degree of protection afforded by the intermolecular interaction to be quantified. The ability to probe the interaction interface within protein–protein and other protein complexes is a particularly powerful application of the RP-MS, which we have pioneered. Not only can one probe the regions of the interface within each interacting protein or peptide molecule, but a quantitative assessment can also be made of the proximity of individual amino acid side chains to one another within the interface. Under the high radical flux and millisecond timescale experiments, reactive amino acid side chains exposed to the solvent over an area greater than 30 $Å^2$ will oxidize and do so to a level that can be directly correlated to its solvent accessibility. An advantage of RP-MS experiments is that the nature of the interaction sites across all binding partners can be analyzed independently.

The key issue in the application of RP-MS for studies of such macromolecular interactions is that the time that the proteins are in complex, and stay in contact, should be longer than exposure of the complex to radicals. When only a proportion of the proteins in solution are in complex (due to their affinity for one another), the level of protection represents rather an average of the population of molecules in solution. Our studies of protein complexes have revealed that RP-MS is applicable to protein complexes with dissociation constants (K_d) at or below the nanomolar range. An initial application of RP-MS to study interaction between profilin, a eukaryotic protein that enhances actin growth, and poly-L-proline, with a dissociation constant of 55 µM, showed that oxidation rates at accessible residue side chains of profilin remained the same (0.57 ± 0.1 s^{-1}) in the presence or absence of the binding peptide poly-L-proline [8]. The interaction of profilin and poly-L-proline is known to reduce the solvent accessibility of the side chains of tryptophan at positions 3 and 31 and tyrosine at positions 6 and 24 by NMR spectroscopy. Synchrotron based RP-MS studies [24], however, showed no significant change in oxidation rates of these residues due to low affinity binding of the complex.

In contrast, the interaction of cytoskeletal protein actin with gelsolin segment-1 (S1) with a dissociation constant in the nanomolar range could be explored by RP-MS [9, 25]. Oxidation rates for two peptides of gelsolin, one in the binding pocket with actin and the other in a nonbinding domain, were examined. The binding peptide of gelsolin-S1 contains residues 96–108, which includes Phe-104 residue, and has an oxidation rate that, in the presence of actin, is reduced by a factor of 35 from 0.57 ± 0.1 s^{-1} to 0.02 ± 0.01 s^{-1}. Oxidation rates for an onbinding peptide containing residues 66–83 of gelsolin-S1 remained unaffected. Applications of RP-MS using an electrospray ionization discharge source for studies of protein interactions were first reported in early 2003 for the ribonuclease S–protein–peptide complex [26]. The complex is formed following the specific cleavage of ribonuclease A with the pro tease s ubstilin. T he 20 res idue S- peptide f rom t he N– terminus reassociates to t he re mainder o f t he pro tein to f orm t he act ive S- complex. There a re t hree o xidizable a mino aci ds wi thin S- peptide (Phe-8, H is-12, Met-13) that all become less accessible to solvent upon binding of S-protein, with the total s ide c hain s urface acces sible a rea mea sured de creasing from 283.1 to 3 3.9Å^2. T o de monstrate t he s pecific i nteraction a nd b inding o f S-peptide with S- protein i n a co mpetitive a ssay e xperiment, t wo s olutions also containing four nonbinding peptides of a similar size to S-peptide were prepared at a pH of 5.5 and 2. The relative abundances of the oxidized forms of the S-peptide ions $[M + 3H + O]^{3+}$ and $[M + 2H + O]^{2+}$ were found to reduce dramatically at pH 5.5, consistent with the protection of its reactive residues upon binding to S- protein, while the level of oxidation at t he other four peptides remained unchanged (Figure 6.9).

Tandem mass spectrometry revealed that oxidation occurs at all oxidizable residues o f S- peptide (Phe-8, H is-12, M et-13) wi th t he majority o f o xygen incorporated at Met-13 under both pH conditions. Segments of ribonuclease S-protein were a lso found to b e protected from oxidation upon formation of the S- complex at p H 5.5. T he pro tection w as l ocalized to res idues 96 –100, over t he ne ighboring res idues 8 5–95, co nsistent wi th t he S- protein b inding domain. Both segments contain common reactive residues in histidine, proline,

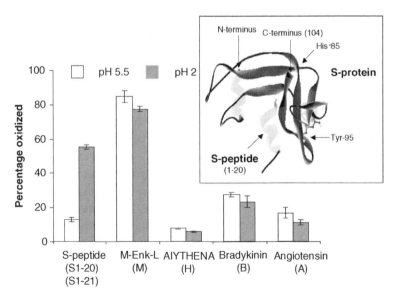

Figure 6.9   Bar g raph representation of the decrease in oxidation levels in r ibonuclease S-peptide and other nonbinding peptides after exposure of an equimolar mixture with S-protein at a pH of 5.5 and 2. Inset shows the X-ray crystal structure of the S-peptide–protein complex.

and e ither p henylalanine o r ty rosine, s uch th at th e r eactivity o f e ach w ith hydroxyl radicals in the absence of any protection would be comparable.

Similar e xperiments w ere co nducted to s tudy t he ca lmodulin–melittin protein–peptide co mplex [27]. U nlike t hat f or t he r ibonuclease co mplex, no h igh-resolution s tructure o f t he ca lmodulin–melittin co mplex ha s b een reported. Models of the complex [28], however, have been proposed. In the presence o f ca lcium, ca lmodulin u ndergoes a s ignificant conformational change with the protein enveloping the 26 res idue polypeptide a xis to form a horseshoe-like structure such that each terminus of the protein is in close contact with the peptide [28].

An e quimolar m ixture o f t hree co mparably s ized (~3 k Da) p eptides, including melittin, was treated with calmodulin in the presence of calcium at pH 4 and 7. The level of oxidation within the nonbinding peptides' insulin A chain and adrenocorticotropic hormone (ACTH, res idues 18–39) was found to be constant at either pH. In melittin, however, it was found to be reduced by 36% at pH 7, consistent with the protection of its reactive side chains upon binding to calmodulin. This reduction in oxidation was localized to the side chain of the tryptophan residue at position 19 in the peptide [27].

A re duction in the oxidation levels across the structure of calmodulin as also followed after digestion of the oxidized protein with trypsin. The levels of oxidation within the peptides were quantified based on the areas under the selected i on c hromatogram f or eac h p eptide i n i ts u noxidized a nd m ono-oxidized forms. A significant degree of protection from oxidation (over 60%) at reactive residue side chains within residues 14–21 and 95–106 of calmodulin is observed when the protein is oxidized in the presence of mellitin, while far less protection is measured elsewhere in the protein. This protection was localized to the phenylalanine residues at positions 16 and 19 in the N-terminal end of the protein, and using tandem mass spectrometry, methionine residues at pos itions 1 44 a nd 1 45 toward t he C t erminus [27]. T hese res ults a re consistent with the horseshoe model for the complex.

Other synchrotron radiolysis-based studies investigated the divalent cation-dependent b inding of act in with gelsolin se gment-1 [29]. Ox idation rat es of binding peptides were found to vary dramatically from 0.57 ± 0.06 s^{-1} to 44 ± 5 s^{-1} within subdomains of actin. Oxidation rates as a function of Ca^{2+} and Mg^{2+} ions were examined for 28 peptides, and rates for several peptides were reduced u pon M g^{2+} ad dition, s ignifying po ssible s tructural reo rganization within subdomains 2 and 4 and a C terminus near subdomain 1. The calcium-dependent activation of gelsolin [30] was also studied. The solvent accessibility as a function of Ca^{2+} concentration for subdomains S1, S2, S4, and S6 were monitored. The stability (K_d) of these domains ranged from 60.4 ± 17.8 (peptide 162–166 s ubdomain S 2) t o 9 7.3 $\pm$ 5. 2 μM (peptide 4 9–72 s ubdomain S 1) measured at the m idpoint of each transition. These results provided evidence of a three-state Ca^{2+}-induced activation process of the gelsolin subdomains from a calcium ion free zone to an intermediate at 10 μM Ca^{2+} and finally to the fully activated form at a saturated concentration of calcium.

The impact of protein oxidation on the interaction behavior has also been assessed by RP-MS. The prolonged oxidation of proteins can promote changes within their tertiary structures and contacts. RP-MS experiments following oxidation over extended (>50 ms) timescales enable one to deduce the impact of oxidation on protein structures and the identification of residues that are critical in stabilizing protein interactions.

6.6   ONSET OF OXIDATIVE DAMAGE AND ITS IMPLICATIONS FOR PROTEIN INTERACTIONS

Protein oxidation has been implicated with a wide range of human diseases and the aging process [31, 32]. Reactive oxygen species (ROS), such as the hydroxyl radical, contribute to this irreversible structural damage and the consequential loss of protein activity and function. It is then of fundamental interest to study the nature and onset of this damage with a view to either halting or slowing the progression of oxidative damage.

It has been illustrated earlier in this chapter that the limited oxidation of proteins on short millisecond timescales (maximum 30–50% for whole protein) is required to prevent protein damage from both a conformational and structural perspective. For apomyoglobin, the onset of the photochemical oxidative damage occurs at approximately 50 ms (Figure 6.5) when oxidation levels begin to exceed this. When the data shown in Figure 6.5 is plotted on a timescale as a function of total oxidation levels, the extent of oxidation is observed to rise and then fall [11]. The fall is associated with the increasing amounts of cross-linked and degraded products formed, which results in less and less oxidized intact protein.

The onset of oxidative damage can be followed within proteins as a whole and/or within particular protein segments after digestion of the exposed protein. This has been illustrated for the protein α-crystallin in which increasing levels of protein oxidation were achieved within an electrical discharge source by reducing the flow rate of the protein solution to the needle tip [33]. α-Crystallin is the most abundant protein within the mammalian eye lens and plays an essential role in maintaining the transparency of the lens and vision [34]. When oxidized, through aging and oxidative damage induced by UV-light, the protein is degraded and its cross-linked forms cause the lens to become opaque. Untreated, such damage can result in the formation of a cataract.

In the first demonstration of RP-MS to study the onset of protein damage [33], different regions of the protein α-crystallin were found to exhibit different

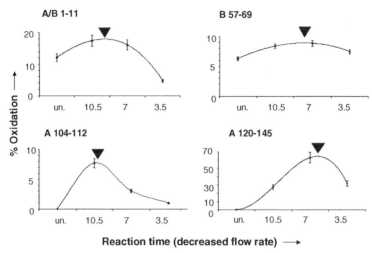

Figure 6.10 Ox idation profiles for segments across the protein α-crystallin after exposure of the protein to hydroxyl radicals over increasing reaction times (lower protein flow rates) [33].

susceptibilities a nd o nsets o f o xidative d amage (Figure 6 .10). N -terminal domains comprising the first 11 residues of the A and B subunits were found to be more susceptible to oxidative damage than residues 120–145 toward the C terminus of the A s ubunit. Elsewhere, residues 104–112 of the A s ubunit were particularly susceptible to damage and were found to oxidize up to only 7% before the onset of damage. In contrast, central residues 57–69 of the B subunit were found to resist both oxidation (with oxidation levels maintained at 7% regardless of the reaction time) and the onset of oxidative damage even at high oxidizing conditions under which the whole protein was found to be appreciably degraded [33]. The data was largely in accord with that predicted in the context of a model structure for the α-crystallin protein.

In the lens of the eye and elsewhere in the body, chaperones and other proteins interact with one another to he lp evade oxidative damage. These interactions, and the protection they confer in terms of oxidative damage, can also be studied by RP-MS. A recent report [35] has examined the protection α-crystallin confers on the taxon-specific upsilon(υ)-crystallin by following the onset of oxidation in segments of the latter protein in the absence and presence of α-crystallin at a 1 :2 ratio. Native gel electrophoresis was used to confirm the formation of the complex with oxidation within υ-crystallin followed at increasing exposure times across eight segments of the protein. Significant protection from oxidation within υ-crystallin, across four of these segments was observed in the presence of α-crystallin, where α-crystallin was found to both reduce their oxidation and delay the onset of oxidative damage. The greatest protection was observed at residues 132–148 and particularly in the vicinity of the side chain of Trp-247 [35].

6.7   APPLICATION OF RADICAL OXIDATION TO STUDY PROTEIN ASSEMBLIES

Prolonged exposure of proteins with reactive oxygen species (ROS) results in the formation of oxidized and misfolded protein, cross-linked protein, and protein aggregates that have been implicated in the genesis of many human diseases. The amyloidoses are a group of protein misfolding diseases including Alzheimer's disease (AD) that result from the extracellular deposition of pathologic insoluble protein fibrils in organs and tissues.

Transthyretin (TTR) is a soluble human plasma protein, and several of its variants are prone to fibril formation. The oxidation of the amino acid side chains of TTR can cause changes within tertiary contacts which promote fibril formation. The effects of oxidation were investigated by comparing the kinetics of fibril formation for the unoxidized and oxidized proteins [36]. The wild-type (WT) and a V30M TTR mutant were reacted with reactive oxygen species (ROS) over extended reaction timescales from several minutes to hours to effect their complete oxidation. The oxidized proteins retained their tetrameric structures as determined by glutaraldehyde cross-linking experiments. Side chain modification of methionine residues at positions 13 and 30 were the dominant oxidative products, with both mono and di-oxidized methionine residues identified by mass spectrometry analysis.

The *in vitro* kinetics of fibril formation was then investigated for both the unoxidized and oxidized proteins. The growth of fibrils for WT-TTR,

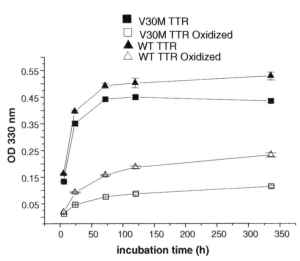

Figure 6.11   Measurements of optical density at 330 nm in solutions of unoxidized and oxidized wild-type (WT) transthyretin and a V30M mutant demonstrating differences in the rate of fibril growth [36].

V30M-TTR, and their oxidized counterparts were individually monitored by turbidity measurements at 330 and 400 nm for up to 14 days (Figure 6.11). The percentage of fibril formation was calculated from the ratios of OD of oxidized and unoxidized proteins (% fibrils = (OD_{330nm} oxidized / OD_{330nm} unoxidized) × 100).

Oxidation had a dramatic affect on initial rates (from the slope of the lines up to 24 hours) of fibril growth for both proteins (Figure 6.11). In the case of WT-TTR, oxidation inhibited the fibril growth by approximately 76% and for V30M-TTR by nearly 90%. Oxidation affected the kinetics of fibril formation for V30M-TTR more than WT-TTR, consistent with the fact the V30M-TTR contains one more methionine residue available for oxidation. These inhibiting effects of oxidation on the fibril growth suggest that domains neighboring the methionine residues are critical in stabilizing the tetrameric and folded monomer structures.

6.8 MODELING PROTEIN COMPLEXES WITH DATA FROM RP-MS EXPERIMENTS

Like other chemical labeling approaches, some of which are described elsewhere in this volume, RP-MS generates significant amounts of solvent accessibility data that needs to be interpreted in the context of protein structures. To enable structures for protein complexes to be proposed from those for their component molecules using RP-MS data, a new docking algorithm known as PROXIMO has recently been developed [37]. The algorithm uses a Katchalski-Katzir geometric fitting routine to dock proteins and score those complexes in terms of the best correlation between a measure of the solvent accessibility surface (SAS) at reactive residues and oxidation levels measured by RP-MS. The 2000 conformations (numbered 0–1999) that demonstrate the greatest surface shape complementarity, as evaluated by the geometric matching algorithm, are scored by PROXIMO based on the correlation of SAS shielding with the oxidation data. Differences in the levels of oxidation at reactive residues in the proteins alone and in complex, together with a geometric fitting routine, are used to assemble and score the structures for proposed protein complexes. A simple to use and intuitive graphical user interface (GUI) is employed to aid data entry and the analysis of the results (Figure 6.12).

A successful illustration of the performance of the algorithm has been reported for the RP-MS described previously for the ribonuclease S-complex. The top five complexes scored by PROXIMO (#157, 488, 113, 1175, and 217) had structures that deviated from the experimentally determined structure (crystal) with RMSD values between 0.45 and 1.26 Å^2 (Figure 6.12). These

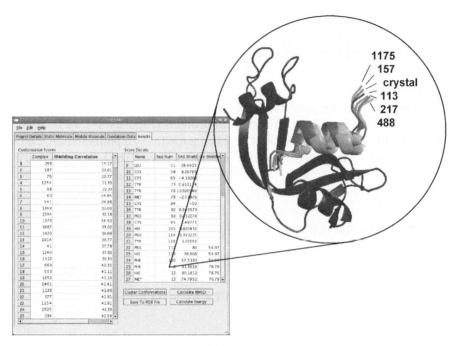

Figure 6.12 T op five scored complexes for the ribonuclease S-complex based on a correlation between calculated solvent accessibilities of amino acid residue side chains and oxidation data obtained from RP-MS experiments generated by the PROXIMO algorithm [37].

values a re a ll within the res olution o btained for the experimental structure (2.1 Å). T he abse nce o f t he R P-MS d ata res ulted i n s tructures i n w hich S-peptide was far removed from its known binding site on S-protein.

6.9 CONCLUSIONS

Rapid Probe-Mass S pectrometry, pioneered i n the late 1990s [7, 8, 12, 19], has a lready de veloped i nto a po werful ne w a pproach wi th w hich to s tudy protein structures, the dynamics of protein folding events, the interactions of proteins, and the onset of oxidative damage of importance in the pathogenesis of disease and aging. Important experimental conditions relating to the time of radical exposure and the extent of protein oxidation must be observed in order to pre vent pro tein d amage w hen s tudying ph ysiologically s ignificant structures and interactions. Extended reaction timescale experiments enable oxidative damage onsets to be explored, both in terms of whole protein and specific pro tein d omains. As wi th t he o ther a pproaches descr ibed i n t his book, it is certain to expand our knowledge on the interactions of proteins and their es sential ro le i n me diating b iological f unctions a nd pro cesses. I n common wi th t he o ther me thods descr ibed i n ea rlier c hapters, ma ss spectrometry is central to the approach, both from the perspective of effecting radical-induced o xidation b y e lectrical d ischarge wi thin a n e lectrospray source, a nd i n t he a nalysis o f o xidized pro tein a nd t heir d igests wi th unprecedented speed, sensitivity, and accuracy. New computer algorithms are assisting with the analysis of the data in terms of modeling protein complexes that have yet to be revealed by other analytical and spectroscopic means.

REFERENCES

1. Tullius, T. D., and Dombroski, B. A. (1986). Hydroxyl radical "footprinting": high-resolution information about DNA–protein contacts and application to la mbda repressor and Cro protein. *Proc. Natl. Acad. Sci. U.S.A.* **83**: 5469–5473.

2. Heyduk, E ., a nd He yduk, T . (1994). Mappi ng prot ein do mains i nvolved i n macromolecular interactions: a novel protein footprinting approach. *Biochemistry* **33**: 9643–9650.

3. Fenton, H. J . H. (1894). O xidation o f t artaric a cid i n t he presenc e o f i ron. *J. Chem. Soc.* **65**: 899.

4. Pierre, J. L. L., and Fontecave, M. (1999). Iron and activated oxygen species in biology: the basic chemistry. *BioMetals* **12**: 195–199.

5. Sclavi, B., S ullivan, M., C hance, M. R ., B renowitz, M., a nd Woodson, S. A. (1998). R NA folding at m illisecond i ntervals by s ynchrotron hydroxyl r adical footprinting. *Science* **279**: 1940–1943.

6. Sclavi, B., Woodson, S., Sullivan, M., Chance, M., and Brenowitz, M. (1998). Following the folding of RNA with time-resolved synchrotron X-ray footprint-ing. *Methods Enzymol.* **295**: 379–402.

7. Maleknia, S. D., and C hance, M. R. (1998). S ynchrotron r adiolysis s tudies of peptides by mass spectrometry. In *Proceedings of the 46th ASMS Conference on Mass Spectrometry and Allied Topics*, p. 972, Orlando, Florida.

8. Maleknia, S . D ., G oldsmith, S ., V orobiev, S ., A lmo, S . A ., C hance, M. R . , and D ownard, K . M. (1998). E xamining prot ein–peptide i nteractions t hrough synchrotron X -ray footprinting t echniques. I n *Proceedings of the 46th ASMS Conference on Mass Spectrometry and Allied Topics*, p. 973, Orlando, Florida.

9. Maleknia, S. D., and Downard, K. M. (2001). Radical approaches to probe protein structure, folding, and interactions by mass spectrometry. *Mass Spectrom. Rev.* **20**: 388–401.

10. Maleknia, S . D ., a nd D ownard, K . M. (2 001). U nfolding o f ap omyoglobin helices by synchrotron radiolysis and mass spectrometry. *Eur. J. Biochem.* **268**: 5578–5588.

11. Maleknia, S. D., Wong, J. W. H., and Downard, K. M. (2 004). Photochemical and e lectrophysical p roduction o f r adicals on m illisecond t imescales to probe

the s tructure, d ynamics a nd i nteractions o f prot eins. *Photochem. Photobiol. Sci.* **3**: 741–748.

12. Maleknia, S. D., Brenowitz, M., and Chance, M. R. (1999). Millisecond radiolytic modification of peptides by synchrotron X-rays identified by mass spectrometry. *Anal. Chem.* **71**: 3965–3973.

13. Buxton, G. V., Greenstock, C. L., Helman, W. P., and Ross, A. B. (1988). Critical review of rate constants for reactions of hydrated electrons, hydrogen atoms and hydroxyl radicals in aqueous solution. *J. Phys. Chem. Ref Data* **270**: 513–886.

14. Masuda, T., Nakano, S., and Kondo, M. (1973). Rate constants for the reaction of OH radicals with enzyme proteins as determined by the *p*-nitrodimethylanaline method. *J. Radiat. Res. (Japan)* **14**: 339–345.

15. Xu, G., and Chance, M. R. (2004). Radiolytic modification of acidic amino acid residues in peptides: probes for examining prot ein–protein i nteractions. *Anal. Chem.* **76**: 1213–1221.

16. Xu, G., and Chance, M. R. (2 005). R adiolytic mo dification a nd r eactivity of amino acid residues serving as structural probes for protein footprinting. *Anal. Chem.* **77**: 4549–4555.

17. Takamoto, K., and Chance, M. R. (2 006). Radiolytic protein footprinting with mass spectrometry to probe the structure of macromolecular complexes. *Annu. Rev. Biophys. Biomol. Struct.* **35**: 251–276.

18. Sharp, J. S., Becker, J. M., and Hettich, R. L. (2003). Protein surface mapping by chemical oxidation: structural analysis by mass spectrometry. *Anal. Biochem.* **313**: 216–225.

19. Maleknia, S. D., C hance, M. R. , a nd D ownard, K . M. (1999). El ectrospray-assisted m odification o f prot eins: a r adical prob e o f prot ein s tructure. *Rapid Commun. Mass Spectrom.* **13**: 2352–2358.

20. Kiselar, J. G., Ma leknia, S. D., S ullivan, M. , D ownard, K . M. , a nd C hance, M. R. (2002). Hydroxyl radical probe of protein surfaces using synchrotron X-ray radiolysis and mass spectrometry. *Int. J. Radiat. Biol.* **78**: 101–114.

21. Thomas, M. C. , M itchell, T. W., a nd B lanksby, S . J . (2 006). Ozon olysis o f phospholipid dou ble b onds d uring e lectrospray i onization: a ne w to ol f or structure determination. *J. Am. Chem. Soc.* **128**: 58–59.

22. Hambly, D. M., and Gross, M. L. (2005). Laser flash photolysis of hydrogen per-oxide to oxidize protein solvent-accessible residues on t he m icrosecond times-cale *J. Am. Soc. Mass. Spectrom.* **16**: 2057–2063.

23. Hughson, F. M., Wright, P. E., and Baldwin, R. L. (1990). Structural characterization of a partly folded apomyoglobin intermediate. *Science* **249**: 1544–1548.

24. Goldsmith, S., Ma leknia, S. D., A lmo, S. C., and Chance, M. R. (1999). Synchrotron X-ray footprinting of profilin poly-proline peptide complex. *Biophysical J.* **76**: A172.

25. Goldsmith, S. C., Ma leknia, S. D., A lmo, S. C., a nd C hance, M. R. (2 000). Synchrotron X -ray f ootprinting o f t he g elsolin–actin co mplex. *Biophysical J.* **78**: 213.

26. Wong, J . W ., Ma leknia, S . D ., a nd D ownard, K . M. (2 003). S tudy o f t he ribonuclease-S-proteinpeptide complex using a r adical probe and electrospray ionization mass spectrometry. *Anal. Chem.* **75**: 1557–1563.

27. Wong, J. W., Maleknia, S. D., and Downard, K. M. (2 005). Hydroxyl r adical probe of the calmodulin–melittin complex interface by electrospray ionization mass spectrometry. *J. Am. Soc. Mass Spectrom.* **16**: 225–233.

28. Scaloni, A., M iraglia, N., O rru, S ., A modeo, P., M otta, A., Ma rino, G., a nd Pucci, P. (1998). Topology of the calmodulin–melittin complex. *J. Mol. Biol.* **277**: 945–958.

29. Guan, J . Q ., A lmo, S . C ., R eisler, E . a nd C hance, M. R. (2 003). S tructural reorganization of proteins revealed by radiolysis and mass spectrometry: G-actin solution structure is divalent cation dependent. *Biochemistry* **42**: 11992–12000.

30. Kiselar, J. G., Janmey, P. A., Almo, S. C., and Chance, M. R., (2003). Visualizing the C a^{2+}-dependent a ctivation o f g elsolin b y u sing s ynchrotron f ootprinting. *Proc. Natl. Acad. Sci. U.S.A* **100**: 3942–3947.

31. Stadtman, E. R. (1992). Protein oxidation and aging. *Science* **257**: 1220–1224.

32. Berlett, B. S., and Stadtman, E. R. (1997). Protein oxidation in aging, disease, and oxidative stress. *J. Biol. Chem.* **272**: 20313–20316.

33. Shum, W. -K., Maleknia, S. D., and Downard, K. M. (2005). Onset of oxidative damage in α-crystallin b y r adical prob e m ass sp ectrometry. *Anal. Biochem.* **344**: 247–256.

34. Horwitz, J. (2003). Alpha-crystallin. *Exp. Eye Res.* **76**: 145–153.

35. Issa, S., a nd Downard, K . M. (2 006). I nteraction b etween a lpha a nd u psilon-crystallin common to the eye of the Australian platypus by radical probe mass spectrometry. *J. Mass Spectrom.*, **41**: 1298–1303.

36. Maleknia, S. D., Reixach, N., and Buxbaum, J. N. (2006). Protein oxidation inhibits amyloid fibril formation of transthyretin. *FEBS J.* **273**: 5400–5406.

37. Gerega, S . K ., a nd D ownard, K . M. (2 006). P ROXIMO — a do cking ne w algorithm to model protein complexes using data from radical probe mass spectrometry. *Bioinformatics* **22**: 1702–1709.

INDEX

Aerolysin, 29, 30
Affinity chromatography, 2
Albumin, 16
Alcohol dehydrogenase, 5, 29, 32
Alzheimer's disease, 128
Average cross section, 12

Binding constant, 5, 16, 52
Biomolecular interaction analysis (BIA),
   38
Bottom–up approach
   in cross-linking MS, 85
   with ESI-MS, 94, 95
   with MALDI-MS, 94
Bottom-up sequencing, 3. *See also* MS/MS

Calmodulin, 69, 70, 95–97, 125
Cancer, 37, 55
Carbonic anhydrase, 17
Collisionally activated dissociation (CAD),
   16–18
Computer software
   for analysis of cross-linking MS data,
      99
   for analysis of RP-MS data, 129

Cross-linkers
   amine-reactive, 89
   heterobifunctional, 93
   homobifunctional, 92
   photoreactive, 91
   sulfhydryl-reactive, 91
   trifunctional, 93
   zero-length, 93
Cryoelectron microscopy, 2
Crystallin, 126, 127

Denaturants, MALDI tolerance, 26
Detergents, 15
Deuterium oxide (D_2O), 46
Dialysis, 15
Dipole–dipole, 26
Dissociation constants, 16, 52

Electron capture dissociation
   (ECD), 88
Electrospray ionization mass
   spectrometry (ESI-MS)
   critical experimental parameters, 8
   of protein complexes, 3
   types of interactions probed, 4

Mass Spectrometry of Protein Interactions Edited by Kevin M. Downard
Copyright © 2007 John Wiley & Sons, Inc.

Fast atom bombardment (FAB), 3, 34
First shot phenomenon, 25, 28, 29
Fourier transform ion cyclotron resonance
  (FT-ICR), 1, 3, 10, 12, 16, 17

Gas-phase electrophoretic mobility
  molecular analyzer (GEMMA), 13, 14
Gas phase dissociation, of protein
  complexes, 29

Hemolysin, 29
Human immunodeficiency virus (HIV), 4,
  6, 15, 20, 21
Human plasma, 38, 39
Hydrogen bonds, 26
Hydrogen exchange (HX), 46–51, 57.
  See also Hydrogen exchange mass
  spectrometry
Hydrogen exchange mass spectrometry
  (HX-MS)
  protein-peptide interactions, 54
  protein-protein interactions, 52
  protein–small molecule interactions, 55
Hydrophobic interactions, 5–8, 26
Hydroxyl radical, 110, 111, 113–119, 121,
  126, 127

Immunoaffinity, 37
Immunoprecipitation, 38
Influenza virus, 35, 36
Infrared (IR) MALDI, 34
Infrared multiphoton dissociation
  (IRMPD), 17, 87, 88, 95
Ion mobility, 1, 3, 5, 12, 13, 15. *See also* Ion
  mobility spectrometry
Ion mobility spectrometry (IMS), 12, 13

Laser fluence, 25, 32, 33
Laser pulse, 28, 34
Liquid matrices, MALDI, 34
Lysozyme, 4, 12, 35

MALDI matrices, 31
MALDI-MS
  on affinity targets and surfaces, 37
  first glimpses of protein complexes, 28
  of protein complexes on conventional
    targets, 35
MALDI-TOF, 38

Matrix criteria, to preserve protein
  complexes, 30
Matrix-assisted laser desorption ionization
  mass spectrometry, *see* MALDI-MS
Mechanisms, for MALDI ionization, 27
Monoclonal antibody, 35–37
MS/MS, 3, 17, 18. *See also* Tandem mass
  spectrometry
Myoglobin (also apomoglobin), 4, 11, 112,
  115, 116, 121, 122, 126

Nanoelectrospray, 9, 15
Native gel electrophoresis, 2
Nickel powder, as MALDI
  matrix, 27
Nobel Prize, 2, 27
Nuclear magnetic resonance (NMR), 2, 7,
  63–65, 70, 71, 73, 78, 84, 95, 96

OrbiTrap, 3, 12
Oxidative damage onsets, application of
  RP-MS, 126

Photolytic damage, 91
PLIMSTEX method, 52
Plume, 27, 35
Porin, 33
Proteasome, 11, 14, 18
Protein assemblies, 65, 88
  studied by RP-MS, 128
Protein–ligand complex, 5, 9, 17
Protein–ligand interactions
  applications of HX-MS, 55
  limited proteolysis of, 74
Protein–nucleic acid interactions, limited
  proteolysis of, 72
Protein–peptide interactions
  applications of HX-MS, 54
  applications of RP-MS, 122
Protein–protein interactions
  applications of HX-MS, 52
  applications of RP-MS, 122
  limited proteolysis of, 69
PROXIMO, algorithm, 129, 130

Quadrupole, 5, 6, 10, 11
Quadrupole ion trap, 94
Quadrupole time-of-flight (QTOF), 5–8, 10,
  11, 17

Radical Probe-Mass Spectrometry (RP-MS)
  genesis of, 110
  important experimental conditions for, 115
  reactive residue side chains in, 111
Radicals, generation of, on millisecond
  timescales, 117
Ribonuclease S (RNase S), peptide, protein,
  5, 20, 124, 129, 130
Ribosome, 2, 16

Site-directed mutagenesis, 57
Size-exclusion chromatography, 86
Stored waveform inverse Fourier transform
  (SWIFT), 95
Streptavidin, 17, 28, 29
Surface-enhanced affinity capture (SEAC),
  37. *See also* Surface enhanced laser
  desorption ionization
Surface enhanced laser desorption
  ionization (SELDI), 37, 38
Surface plasmon resonance (SPR), 2, 38
Sustained off-resonance irradiation
  collision-induced dissociation
  (SORI-CID), 95

Synchrotron light, 110, 111, 117

Tandem mass spectrometry, 1, 3, 10, 16, 17.
  See also MS/MS
Tandem TOF (also TOF/TOF), 94
Time-of-flight (TOF), 3, 6, 10, 11, 16
Top-down approach, in cross-linking MS,
  88
Top-down sequencing, 3
Transthyretin, 128

Ultracentrifugation, 2
Ultraviolet (UV) MALDI, 34

Van der Waals forces, interactions, 6, 26
Vancomycin, 16

Wavelength, laser, 25, 27, 32–34

X-ray crystallography, 2, 3, 14, 63, 64, 71,
  76, 78, 84

Yeast two-hybrid assay, 2

Printed and bound by CPI Group (UK) Ltd, Croydon, CR0 4YY

27/10/2024

14580256-0005